Biogeochemical Cycles

The Computer-Based Earth System Science Series
William L. Chameides, editor

Biogeochemical Cycles

A Computer-Interactive Study
of Earth System Science
and Global Change

W.L. Chameides
E.M. Perdue

New York   Oxford
Oxford University Press
1997

OXFORD UNIVERSITY PRESS

Oxford New York
Athens Auckland Bangkok Bogotá Bombay
Buenos Aires Calcutta Cape Town Dar es Salaam Delhi
Florence Hong Kong Istanbul Karachi
Kuala Lumpur Madras Madrid Melbourne
Mexico City Nairobi Paris Singapore
Taipei Tokyo Toronto

and associated companies in

Berlin Ibadan

Library of Congress Cataloging-in-Publication Data
Chameides, W. L.
Biogeochemical cycles : a computer-interactive study of earth
system science and global change / W.L. Chameides, E.M. Perdue.
p.   cm. — (The Computer-based earth system science series)
Includes bibliographical references and index.
ISBN 0-19-509279-1 (cloth)
1. Biogeochemical cycles. 2. Biogeochemical cycles—Mathematical
models. I. Perdue, Edward M. II. Title III. Series.
QH344.C45   1996
577'.14—dc21   96-38035
                                                                      CIP

1   3   5   7   9   8   6   4   2
Printed in the United States of America
on acid-free paper

Contents

Preface

Global biogeochemical cycles lie at the intellectual core of the scientific discipline that has come to be known as *Earth System Science*. The vast and complex array of biological, geological, and chemical processes that comprise these cycles transform and transport the elements through the various components or spheres of our planet, keeping the earth's chemical system in working order and determining the gross chemical and physical properties of our global environment. For this reason, anyone interested in understanding the causes of global environmental change and its implications for life on earth would be well-advised to begin with an investigation of global biogeochemistry.

However, it is also true that the undertaking of a course in global biogeochemical cycles presents significant challenges for the student and the teacher. As disciplinary scientists, biogeochemical cycles force us to be interdisciplinary; the physicist must be willing to delve into chemistry and biology, while the biologist must learn physics as well as chemistry, and so forth. As specialists in the earth sciences, biogeochemical cycles force us to be holistic; for example, the atmospheric scientist must concern him- or herself with ocean sciences, geophysics, ecology, and so on. The study of biogeochemical cycles can even confront us from time to time with the metaphysical and metaphorical—for example, when we consider James Lovelock's Gaia Hypothesis[1] in which global biogeochemical cycles are likened to the metabolism of a living planetary organism.

Mathematics and numerical modeling provide an additional challenge for the student and teacher of biogeochemical cycles. Although the unraveling of the wondrous maze of biological and abiological processes that comprise biogeochemical cycles can be a fascinating intellectual exercise, it can only be a qualitative exercise in the absence of mathematical and numerical models. The application of mathematical and numerical models transforms this exercise into a quantitative analysis of the dynamics of biogeochemical cycles, their propensity for fostering global change, and their role in determining the past and potential future course of planet earth.

This book represents an attempt to combine all of these important elements of the study of biogeochemical cycles into an integrated and comprehensive text. To-

---

[1]The Gaia Hypothesis, eloquently enunciated in James Lovelock's book entitled *Gaia: A New Look at Life on Earth* (Oxford University Press, 1979), proposes that the earth can be viewed as a living organism that acts to manipulate the chemical and physical environment for the benefit and maintenance of life on earth. The hypothesis contrasts sharply with the more accepted Darwinian view that sees life in competition with itself and striving to adapt to a haphazard environment and largely inanimate planet.

ward this end, Chapters 1, 2, and 3 provide an introduction and review of the fundamentals (i.e., basic chemical concepts, relevant features of the earth system, and the key physical, biological, and chemical processes at work in this system). Depending upon the reader's scientific background and training, portions or all of this material may be skipped. Chapter 4 presents a review of the mathematical formalism used to represent biogeochemical cycles in terms of a system of differential equations and the techniques used to solve these equations. The basic concepts are illustrated with a simple cycle involving humanity and the fictitious "University of Biogeochemistry". As a further aid to the reader, we have developed and included with the book a computer program entitled *BOXES*. The program, introduced at the end of Chapter 4, provides a user-friendly environment for constructing numerical models of biogeochemical cycles without getting bogged down in the detailed numerics of such models. (Students with a minimal knowledge of differential equations and numerical techniques should find it possible to use *BOXES* without fully digesting the material in Chapter 4.) We have found this program to be an invaluable teaching tool; it allows students to work interactively with the professor and also serves as the basis for individual and group projects executed outside of the classroom.

Chapters 5, 6, 7, and 8 contain detailed discussions of the global cycles of P, C, S, and N, respectively. In each case, *BOXES* is used to illustrate the key features of the cycle. Chapter 9 integrates the cycles of P, C, S, and N in order to investigate the stability of atmospheric oxygen, a molecule whose presence on earth is unique within the solar system. Problem sets are included at the end of each chapter; many of these problems can be solved using *BOXES*.

The book has grown out of our experience teaching Global Biogeochemical Cycles at Georgia Tech, and we thank the many graduate students who witnessed our initial attempts to cover this challenging subject and labored through our early versions of *BOXES*. We also owe a special debt of gratitude to Dr. A. C. Lasaga of Yale University, whose seminal work on the "Kinetic Treatment of Geochemical Cycles " (published in *Geochimica et Cosmochimica Acta* in 1980) provided the mathematical formulation for the numerical code in *BOXES*.

W.L. Chameides
E.M. Perdue
Atlanta, GA

Biogeochemical Cycles

Biogeochemical Cycles: Their Role in the Earth System

1

"Biogeochemistry: a science that deals with the relation of earth chemicals to plant and animal life."

Webster's New Collegiate Dictionary

1.1. INTRODUCTION

Virtually every biologically mediated chemical reaction on earth relates in some way to a biogeochemical cycle and ultimately to our species. As members of the earth's biological community, human beings participate in and depend upon these cycles for our very existence. Perhaps the most familiar of the biogeochemical cycles is the one involving the cycling of carbon (C) and oxygen (O) via *respiration* and *photosynthesis* (Figure 1.1). In our homes, for example, house plants, by way of photosynthesis, consume carbon dioxide (CO_2) and produce molecular oxygen (O_2), while we consume O_2 and produce CO_2 via respiration. Of course on a larger scale, photosynthesis and respiration play a much more profound role in supporting life. In addition to producing O_2, photosynthesis allows green plants to store radiant energy from the sun in the form of *chemical energy* by combining water (H_2O) and the carbon from CO_2 into organic molecules such as carbohydrates. This chemical energy is released when respiring organisms assimilate the organic carbon and O_2 produced by green plants and cause them to combine in an exothermic or energy-releasing reaction. In the process, CO_2 and H_2O are returned to the environment and respiring organisms obtain the energy they need to live and grow.

The simple example described above illustrates three key characteristics of a biogeochemical cycle. For one, the cycle describes the chemical and physical transformations of an element on the earth, hence the syllable *geo* in biogeochemical. Secondly, the cycle almost always involves at least one biologically driven or bi-

1

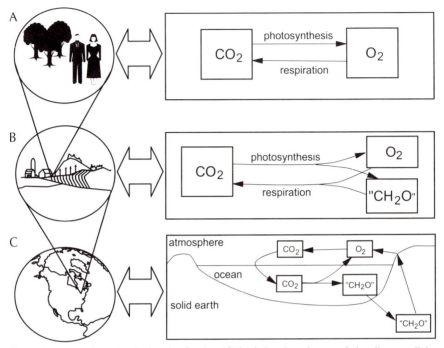

Figure 1.1. The biogeochemical cycle of carbon (C) is vital to the existence of virtually every living organism on earth. **(A)** At its simplest level, we can think of this cycle as comprising two biogeochemical processes: photosynthesis by green plants, leading to the destruction of carbon dioxide (CO_2) and the concomitant production of molecular oxygen (O_2); and respiration by animals, leading to the production of CO_2 and the destruction of (O_2). **(B)** In addition to generating O_2, photosythesis also produces organic compounds (indicated by "CH_2O"). This organic matter is combined with O_2 during respiration to replace the CO_2 lost from the system during photosynthesis. In the process of combining "CH_2O" with O_2, energy is released that can be used by the respiring organism to carry out its metabolic processes and maintain life. **(C)** At the global scale, the cycle of C is considerably more complex and involves the following: (i) dissolution of atmospheric CO_2 into the ocean; photosynthesis by phytoplankton causing the conversion of the dissolved CO_2 to O_2 and "CH_2O"; the transfer of the O_2 to the atmosphere; the sinking and burial at the ocean bottom of a small fraction of the "CH_2O" produced by photosynthesis; the eventual transport over many millions of years of the buried "CH_2O" to the earth's surface by *tectonics* and uplift; and the oxidation of this "CH_2O" via *weathering*, thereby removing the atmospheric O_2 and replacing the CO_2 initially lost from the atmosphere at the beginning of the cycle.

otic process, hence the prefix *bio*. Finally, because chemicals consumed in one process are eventually reformed in a subsequent process, we refer to these processes in the aggregate as *cycles*. And thus we have *biogeochemical cycles*.

1.2. OPEN AND CLOSED CYCLES

Biogeochemical cycles can be "open" or "closed." In an open cycle, material can flow into or out of the cycle. For example, the photosynthesis/respiration cycle of C and O in a house described earlier is an example of an open cycle. Undoubtedly, significant fractions of the O_2 and CO_2 present within any house come from the outside and are consumed by organisms that reside outside. A closed cycle, on the other hand, is a cycle where there is no flow of material into or out of the cycle. In this case we say that the total amount of each element within the cycle is "conserved," much like a physicist would say that the First Law of Thermodynamics is a statement of energy conservation.

To a very close approximation, the earth system (i.e., the ocean, atmosphere, and solid earth) is a closed system with no flow of material out of or into the system. (Small amounts of material do accrete on the earth from meteors, cosmic dust, and so on, and small amounts of mass are lost from the earth as a result of the escape of hydrogen to space, but these represent very small percentages of the total mass of the earth.) Thus, so-called *global* biogeochemical cycles, which describe the cycling of the elements on a global scale and are the main focus of this book, can, in almost every case, be treated as closed cycles. Note, for example, that in the illustration of the carbon cycle in Figure 1.1, only the global cycle shown in part C of the figure is a closed cycle in which every atom of C converted from CO_2 to organic C during photosynthesis is returned to the system as CO_2 at a later point.

Because global biogeochemical cycles are closed cycles, by definition they cause no net chemical change. Since every chemical consumed by one process in the cycle is produced by another, there can be no net production or destruction of any compound over one complete traversal of the cycle. For example, consider our simple photosynthesis/respiration cycle for C and O. While each of these processes is actually comprised of a multitude of individual *elementary reactions*, we can represent the sum total of these reactions as a single *stoichiometric reaction*. Photosynthesis can be represented stoichiometrically as

$$CO_2 + H_2O + h\nu \rightarrow \text{``CH}_2\text{O''} + O_2 \qquad (R1.1)^1$$

while the stoichiometric reaction for respiration is given by

$$\text{``CH}_2\text{O''} + O_2 \rightarrow CO_2 + H_2O \qquad (R1.2)$$

In these reactions we use $h\nu$ to represent the utilization of radiant energy from the sun and use "CH_2O" as shorthand notation to represent organic matter. An organic compound is a compound that contains at least one C atom bonded to at least one H atom. The most common organic compounds in most organisms are carbohydrates. These compounds contain C, H, and O, usually in a stoichiometric ratio of about 1:2:1. Thus "CH_2O" is often used to represent these compounds in simple stoichiometric reactions such as (R1.1) and (R1.2). As we begin to consider more complex systems in subsequent chapters, we will find the need to use a more complete stoichiometric description for photosynthesis and for organic compounds.

The net chemical effect of photosynthesis and respiration can now be inferred by summing (R1.1) and (R1.2). If we do this we obtain a reaction in which every chemical compound that appears on the left-hand side of the reaction also appears on the right-hand side; that is,

$$CO_2 + H_2O + CH_2O + O_2 + h\nu \rightarrow CO_2 + H_2O + CH_2O + O_2 \quad (R1.1 + R1.2)$$

---

[1]The notation (R$x.y$) will be used to denote chemical reactions, where x represents the chapter number where the reaction first appears and y represents the order of appearance in the chapter. Hence, (R1.1) and (R1.2) signify the first and second reactions of Chapter 1, respectively.

And thus we conclude that the sequence of photosynthesis and respiration has no "net chemical effect." While the cycles to be considered in this book will in general be considerably more complex than the simple photosynthesis/respiration cycle considered here, they should, with very few exceptions, have this same basic property; that is, if one sums together all the stoichiometric reactions that make up the cycle, there should be no net chemical change.

1.3. WHY DO WE CARE ABOUT BIOGEOCHEMICAL CYCLES?

Given that biogeochemical cycles do not lead to a net chemical change, one might legitimately ask the question, "Why should we bother to study them?" The answer is that, while biogeochemical cycles do not lead to net chemical changes, they nevertheless fulfill a number of functions that are critically important to the maintenance of life on the earth. One of these functions has to do with the storage and utilization of the radiant energy the earth receives from the sun. Biogeochemical cycles generally describe a pathway by which radiation from the sun is assimilated by living organisms and stored in the form of chemical energy (i.e., the solar energy is used to make compounds that are easily converted into heat through chemical reactions with other materials in the environment). This chemical energy or fuel can be stored by an organism for later use, it can also be transported from one location to another, or exchanged from one organism to another (typically this "exchange" occurs through some form of grazing or predation), until it is eventually consumed for the purpose of supporting the *metabolic* needs of a specific organism within the *biosphere* and releasing heat to the environment.

Another important function is the recycling of material. Because the earth is a closed system, there is a finite amount of material available for use by the biosphere. Global biogeochemical cycles serve as an enormous recycling system that allows the biosphere to use the same elements in its metabolic processes over and over again. In the absence of such cycles, the biosphere would eventually grind to a halt, suffocating in its own wastes. For instance, by way of illustration let us consider the global cycle of C represented in part C of Figure 1.1. Now imagine what would happen if we suddenly "turned off" that portion of the cycle that transports buried organic C from the ocean bottom to the earth's surface. Eventually all of the C now present in the atmosphere and ocean would be converted to "CH_2O" and buried at the bottom of the ocean, there would be no more CO_2 available for photosynthesis, and we would have on our hands a food crisis of major proportions.

Dr. James Lovelock has suggested a useful analogy for understanding the relationship between life on earth and its physical and chemical environment—that is, the analogy of the earth to a single living organism; this is the so-called *Gaia Hypothesis*.[2] Although this hypothesis has been criticized for being overly simplisitic,

---

[2]Initial discussions of the Gaia Hypothesis can be found in Lovelock, J. E., Gaia as seen through the atmosphere, *Atmospheric Environment*, **6**, 579–580, 1972; and Margulis, L., and J. E. Lovelock, Biological modulation of the earth's atmosphere, *Icarus*, **21**, 471–489, 1974. These discussions were followed by two fascinating and eminently readable books by James Lovelock: *Gaia: A New Look at Life on Earth*, Oxford University Press, 1979; and *The Ages of Gaia: A Biography of Our Living Earth*, W. W. Norton and Co., London, 1988.

it nevertheless offers some useful insights. If we were for the moment to adopt this analogy, then global biogeochemical cycles might be viewed as the metabolic pathways of this single large planetary organism. And we, as students of global biogeochemical cycles, might be viewed as "geophysicians," studying the earth's metabolic pathways in much the same way that a medical doctor might take the pulse of a patient or analyze a sample of blood. In fact, just as a medical doctor uses measures of metabolic function to diagnose a patient's health, we will find that global biogeochemical cycles hold the key for our understanding and ultimately predicting global change, whether caused by humanity or by some natural phenomenon.

1.4. WHICH ELEMENTS SHOULD WE STUDY?

Having established that global biogeochemical cycles are important to study, we might next ask ourselves, "Which of the scores of elements that are found on and in the earth should we investigate first?" There is, of course, no single correct answer to this question; the element or elements one studies in biogeochemistry depends to a large extent upon which scientific questions one wants to pursue. In our case, we have argued that we are interested in biogeochemical cycles because of the role they play in defining the interactions between living systems and the chemical and physical environment in which these living systems exist. It seems reasonable then, given this interest, that we focus on the cycles of those elements that have the strongest interaction with living systems. How then do we identify those elements that interact most strongly with living systems? As an opening strategy, we can decide to focus on those elements that are most abundant in living tissue, as they must necessarily cycle into and out of living systems at the greatest gross rate. Analysis of living and dead tissue typically indicates the presence of six major elements: hydrogen (H), carbon (C), oxygen (O), nitrogen (N), phosphorus (P), and sulfur (S), usually in that order of abundance. C, H, O, and N are not surprising since these are the fundamental elements used to make *amino acids*, which in turn are the basic building blocks for all *proteins*. P has two special roles that make it an essential element for living organsims: As a *phosphate ester*, it acts to bind together the individual *nucleotides* that make up a cell's *DNA*; and as *adenosine triphosphate* (or ATP) within a cell's *mitchondria*, it plays a crucial role in the respiratory metabolism of a cell that burns fats and carbohydrates and produces energy that can be used by the cell to do its work. The role of S in living systems can be attributed to its presence in two key two amino acids (cysteine and methionine) that are thought to help provide mechanical structure to the proteins that make up living tissue.

We will focus here on the biogeochemical cycles of five of the six major elements noted above—that is, C, O, N, P, and S. C, N, P, and S are the primary nutrient elements. As we will learn in subsequent chapters, their abundances often affect the productivity and size of ecosystems; ecosystems, in turn, can have a major impact on their abundances and distributions on the earth. For these reasons, the cycles of these elements occupy center stage in the study of biogeochemistry. O is of importance because its abundance determines the *redox* conditions of an environment and, as a result, the types of metabolic pathways an organism can adopt to extract energy from that environment; H, on the other hand, is not addressed here. Because this element is so abundant on the earth and its cycle is so tightly bound to that of C and O, it turns out little can be learned from this cycle that cannot be learned from the cycles of C and O.

And finally, having identified the elements of primary interest for our study of biogeochemical cycles, we need to consider the kinds of compounds or chemical species these elements might form as they participate in these cycles. A partial list-

TABLE 1.1
Some of the Chemical Species Found in Nature Containing C, N, S, and P: Their Relative Oxidation States and Acidities

	Aqueous Solutes							Gases	Solids
	Acidic →	→	→	→	→	→	→ Basic		
A. C Compounds									
Oxidized	H_2CO_3			HCO_3^-		CO_3^{2-}		CO_2	$CaCO_3$
↓								CO	
↓				↑					↑
↓				↑					↑
↓		←	←	←"CH_2O"→	→	→		CH_2O	"CH_2O"
↓				↓					↓
↓				↓					↓
↓								CH_3OH	
Reduced								CH_4	
B. N Compounds									
Oxidized	HNO_3					NO_3^-		N_2O_5, HNO_3	$NaNO_3$
↓								NO_2	
↓	HNO_2					NO_2^-		HNO_2	
↓								NO	
↓								N_2O	
↓								N_2	
Reduced	NH_4^+					NH_3		NH_3	$(NH_4)_2SO_4$
C. S Compounds									
Oxidized	H_2SO_4			HSO_4^-		SO_4^{2-}			$CaSO_4$
↓									
↓	H_2SO_3			HSO_3^-		SO_3^{2-}		SO_2	
↓									FeS_2
Reduced	H_2S			HS^-		S^{2-}		H_2S, $(CH_3)_2S$	FeS
D. P Compounds									
Oxidized	H_3PO_4			$H_2PO_4^-$ $^-HPO_4^{2-}$		PO_4^{-3}			$Ca_5(PO_4)_3OH$
↓									
↓	H_3PO_3			$H_2PO_3^-$		HPO_3^{2-}			
↓									
Reduced								PH_3	

ing of some of the simple compounds the five elements can form is presented in Table 1.1. It is apparent from this listing that the elements can be found in quite a variety of chemical forms, having a wide range of *oxidation states* and *acidities* and existing in different *phases*. What are the processes that determine which of these myriad chemical species will predominate for a given set of environmental conditions? The answer to this question can be found in the basic principles of chemical thermodynamics—the subject of Chapter 2.

SUGGESTED READING

Broecker, W., *How to Build a Habitable Planet*, Eldigo Press, Palisades, New York, 1985
Garrels, R.M., F.T., MacKenzie, and C. Hunt, *Chemical Cycles and the Global Environment*, William Kaufman, Los Altos, Calfornia, 1975
Lovelock, J., *The Ages of Gaia: A Biography of Our Living Earth*. W. W. Norton, New York, 1988.
Sagan, C., *Cosmos*, Random House, New York, 1980.

Principles of Chemical Thermodynamics

2

2.1. INTRODUCTION

Thermodynamics describes the rules that govern the flow and transfer of energy in a system as it makes a transition from one state to another. *Chemical thermodynamics* describes how these rules apply to chemical transformations and the uptake or release of energy that accompany these transformations. In this chapter we review some of the central principles of chemical thermodynamics. This review is not intended to be comprehensive but rather a summary of the essential relationships that we will need in our analysis of the global biogeochemical cycles.

2.2. THE BASICS

All of thermodynamics arises from three basic physical constraints or laws; these are the so-called First, Second, and Third Laws of Thermodynamics. Of principal interest to our discussion are the first two of these laws; in simple terms they state the following:

1. Energy is conserved in any process.
2. The entropy of the universe is increasing.

The First Law—the law of energy conservation—governs the makeup and states of the initial and final substances in any process. It states that the total energy contained in the final state of a system must be equal to the total energy of the initial state of the system (less any energy lost from the system to the surrounding universe). However, the First Law does not place limitations on the direction of the change. For instance, the First Law places no limitation on the direction an object will take in a gravitational field. According to the First Law, a ball can spontaneously "jump" from the ground to a height of 100 feet as long as the ball's initial and final kinetic plus potential energies are equal. Similarly, according to the First Law, a cold body can lose heat to a warm body, just as a warm body can cool by giving up its heat to a cool body. From our experience we know, however, that balls do not spontaneously jump to heights of 100 feet and cool bodies do not become colder by spontaneously transferring their heat to a warmer body. The fact that such events do not occur is covered by the Second Law—the law of increasing entropy—which asserts that isolated systems tend to spontaneously move from more to less ordered states.

The laws of thermodynamics have two principal ramifications for chemical systems (Table 2.1):

TABLE 2.1
Manifestations of the First and Second Laws of Thermodynamics

Process	Result of First Law	Result of Second Law
Motion:		
Ball is released in a gravitational field.	Sum of kinetic plus gravitational potential energy of ball is constant. No constraint on direction of motion—ball can spontaneously jump upward or fall downward.	Ball must fall.
Heat exchange:		
A hot body and a cold body are placed in close contact.	Total energy is constant. No constraint on direction of heat flow. Hot body can heat cold body, and cold body can heat hot body.	Heat will only flow spontaneously from hot body to cold body.
Chemical reaction:		
Species A and B are mixed together and spontaneously react to form species C and D.	The energy released to the environment (i.e., an exothermic reaction) or absorbed from the environment (i.e., an endothermic reaction) is equal to difference between the chemical energy stored in C and D minus that stored in A and B. No constraint on the chemical energy and entropy contained in C and D relative to that of A and B.	Spontaneous reactions minimize chemical energy or enthalpy and maximize entropy. Thus C and D must have lower Gibbs free energy than A and B.

1. From the First Law of Thermodynamics, all reactions must conserve energy.

2. From the Second Law of Thermodynamics, reactions that occur spontaneously (that is, reactions that cannot be reversed without application of heat or work) must result in a net decrease in the chemical potential or so-called *Gibbs free energy* of the system. As we shall see later, a decrease in the Gibbs free energy tends to move a chemical system toward a state where the chemical energy (or enthalpy) stored in the species of that system is minimized while the molecular disorder (or entropy) of the species in that system is maximized.

To illustrate how Gibbs free energy affects the direction of chemical change, we consider a simple chemical system that can shift between two states—one containing *a* moles of molecule A and *b* moles of molecule B, and the other containing *c* moles of molecule C and *d* moles of molecule D. This shift or transition between the two states is represented by a coupled pair of forward and backward elementary reactions:

$$aA + bB \rightleftarrows cC + dD \qquad \text{(R2.1)}$$

Note in (R2.1) that we use a double arrow ($\rightleftarrows$) to represent two reactions, one that proceeds from left to right, and the other that proceeds from right to left. Moreover, while (R2.1) represents two reactions with A and B representing reactants in one and products in the other (and vice versa for C and D), we will adopt convenient shorthand notation: We will refer to A and B as the "reactants" because they are on the left-hand side of (R2.1), and we will refer to C and D as the "products" because they are on the right-hand side of (R2.1).

At equilibrium, there is by definition no net change and thus the rates of the forward and backward reactions must be equal. From the Law of Mass Action it follows that

$$k_{f2.1} \{A\}^a \{B\}^b = k_{b2.1} \{C\}^c \{D\}^d \qquad \text{(2.2.1)}$$

where $k_{f2.1}$ and $k_{b2.1}$ represent the rate constants for the forward and backward reactions of (R2.1) and $\{X\}$ represents the *activity* or effective concentration of species X (see Text Box 2.1). The equilibrium constant, $K_{2.1}$, for (R2.1) is defined as the ratio of the forward and backward rate constants, and thus by rearranging Equation (2.2.1) we obtain the familiar expression for the equilibrium constant in terms of the *reaction quotient:*

$$K_{2.1} = \frac{k_{f2.1}}{k_{b2.1}} = \frac{\{C\}^c \{D\}^d}{\{A\}^a \{B\}^b} \qquad \text{(2.2.2)}$$

Because of the Second Law of Thermodynamics, the value of K, the equilibrium constant, for any given reaction pair is constrained to a specific functional form that depends on the thermochemical properties of the reactants and products and the physical state of the chemical system (e.g., its temperature). This function is given by

$$K = e^{-\Delta G^0 / RT} \qquad \text{(2.2.3)}$$

TEXT BOX 2.1	"ACTIVITY" PROVIDES A THERMOCHEMICALLY CONSISTENT MEASURE OF CONCENTRATION FOR SPECIES IN DIFFERENT PHASES

The activity of a species, {X}, is simply the species' effective concentration. The units used to describe this effective concentration of course depend upon the phase of the species or compound. For instance, the activity of a gas is expressed in terms of its partial pressure, p, in units of atmospheres (atm). Thus for a gas

$$\{X\}_{gas} = p(X) \quad \text{(in units of atmospheres or atm)}$$

Under ideal conditions, the activity of a species dissolved in a liquid water solution (i.e., an *aqueous-phase species*) is simply equal to its concentration. As the concentration of ions in solution increases to levels approaching 1 mole per thousand moles of water, however, *solvation effects* usually cause the solution to diverge from ideality and the activity of an aqueous-phase species in this type of solution can be expected to differ significantly from its concentration. However, for most solutions of interest here, solvation effects tend to be small, and thus we will usually assume that the activity of a dissolved species is equal to its concentration. Units typically adopted for the concentration of a solute include molality (m = moles per kilogram of solvent) and molarity (M = moles per liter of solution). We will use the latter unit—molarity—and thus

$$\{X\}_{aq\text{-}phase} = [X] \quad \text{(in moles/liter, or M)}$$

Finally, species in the solid phase and pure liquids have unit activity and thus

$$\{X\}_{soild} = 1 \quad \text{and} \quad \{X\}_{liq} = 1$$

By convention, we refer to a species in its *standard state* as a species with an activity of unity. Thus pure solids and liquids as well as gases at 1 atm pressure and dissolved species at a concentration of 1 M are all, by definition, in their standard states.

where R ($=1.987$ cal mole^{-1} K^{-1} $= 8.314$ J mole^{-1} K^{-1}) is the *gas constant*, T is temperature in degrees Kelvin (K), and ΔG^0 is the change in the standard *free energy of formation* (or *Gibbs free energy*) in going from the left-hand side of the reaction couplet (i.e., the reactants) to the right-hand side (i.e, the products) of the reaction. In other words,

$$\Delta G^0 = \sum [G^0 \text{ of products}] - \sum [G^0 \text{ of reactants}] \qquad (2.2.4)$$

or, more specifically, for (R2.1)

$$\Delta G^0 = c\Delta G_f^0(C) + d\Delta G_f^0(D) - a\Delta G_f^0(A) - b\Delta G_f^0(B) \qquad (2.2.5)$$

where $\Delta G_f^0(X)$ is the Gibbs free energy needed to form 1 mole of compound X from its elements in their standard states.

In a heuristic sense, we can think of the Gibbs free energy of a compound as the amount of internal or chemical bonding energy minus the entropy (in energy units) associated with the compound and its molecular structure. From this definition we see how the Second Law of Thermodynamics governs chemical equilibria. For instance, suppose the products C and D of (R2.1) have less internal energy and more entropy than those of reactants A and B. It follows then that the Gibbs free energy of C and D will be less than that of A and B and, from Equation (2.2.5), that $\Delta G^0 < 0$. But from Equation (2.2.3), we see that the more negative the value of ΔG^0, the larger K will be and, as a result, the greater the equilibrium will be shifted to the right-hand side in favor of C and D and at the expense of A and B. If on the other hand, the products of (R2.1) have relatively high internal energies and/or low entropies, ΔG^0 will be greater than 0 and the equilibrium will be shifted to the left-hand side of the reaction. Thus we see that Equation (2.2.3) is simply a quantitative statement of the fact that thermodynamics forces chemical systems to move toward states that minimize the internal energy and maximize the entropy of that system (see Text Box 2.2).

TEXT BOX 2.2	SPONTANEOUS REACTIONS: EXOERGIC, BUT NOT NECESSARILY EXOTHERMIC

An *exothermic* reaction is a reaction that results in the net release of heat to the environment, and an endothermic reaction is one where heat is taken from the environment. By the First Law of Thermodynamics, the heat released to or absorbed from the environment must be balanced by a net decrease or increase in the internal energy or enthalpy H, contained in the molecules of the system. Thus, for an exothermic reaction we have $\Delta H < 0$, and for an endothermic reaction we have $\Delta H > 0$.

By contrast, the Second Law of Thermodynamics requires that a spontaneous reaction involves a net decrease in the Gibbs free energy, G, of the system. Thus, $\Delta G < 0$ for spontaneous reactions. Reactions where there is a net decrease in the Gibbs free energy are said to be *exoergic*.

While many spontaneous, or exoergic, reactions are also exothermic, this is not always the case. To understand how this can occur, we must turn to the actual definition of the Gibbs free energy:

$$G = H - TS$$

where S is defined as the entropy and T is temperature. Now consider an isothermal reaction; the change in the Gibbs free energy is given by

$$\Delta G = \Delta H - T\Delta S$$

If the reaction is spontaneous, then we have

$$\Delta G = \Delta H - T\Delta S < 0$$

It thus follows that a spontaneous reaction can also be endothermic (i.e., $\Delta H > 0$), provided that the increase in entropy is large enough to offset the increase in internal energy; that is,

$$\Delta S > \Delta H / T$$

2.3. ACID–BASE EQUILIBRIA

Many of the key biogeochemical pathways in the cycles we shall consider occur in the ocean, and as a result aqueous-phase chemistry (i.e., chemistry of species dissolved in liquid water solutions) will play a critical role in our discussions. Of particular importance are the thermodynamic equilibria that determine the speciation or chemical form of the various compounds that might be present in solution, and key among these equilibria are the ones that relate to acids and bases. Thus our objective in this section is to develop a basic level of understanding of how the chemical form of an acid or base varies with the acidity or pH of the solution.

We adopt here the *Brønsted definition of acids and bases*, where an acid is defined as any substance that can donate hydrogen ions (H^+) in a chemical reaction and a base is defined as any substance that can accept H^+ in a chemical reaction. A species with both acidic and basic properties is referred to as being *amphoteric*. Water itself is an example of an amphoteric species, in that it can donate an H^+, that is,

$$H_2O \leftrightarrows H^+ + OH^- \tag{R2.2}$$

and it can accept an H^+, that is,

$$H_2O + H^+ \leftrightarrows H_2OH^+ \tag{R2.3}$$

Note that while H_2O acts as an acid in the portion of (R2.2) that proceeds from left to right, the hydroxide ion (OH^-) acts as a base in the portion of (R2.2) that proceeds from right to left. In recognition of this fact, we say that OH^- is the *conjugate base* of H_2O. In a similar vein, H_2OH^+ is the *conjugate acid* of H_2O. We define a *polyprotic acid* as an acid that has more than one H^+ to donate; sulfuric acid (H_2SO_4), carbonic acid (H_2CO_3), and phosphoric acid (H_3PO_4) are all examples of polyprotic acids that we will encounter in our discussions of biogeochemical cycles.

We can represent an acid generically by the formula H_nA, where n is the number of acidic hydrogen ions that the acid can potentially donate and A is any species that can exist in the aqueous phase as an anion with n negative charges (i.e., A^{n-}). Thus there are n stepwise equilibria or acid dissociation reactions that can be written to represent the donation of each of these ions:

$$H_nA \leftrightarrows H^+ + H_{n-1}A^- \tag{R2.4a}$$

$$H_{n-1}A^- \leftrightarrows H^+ + H_{n-2}A^{2-} \tag{R2.4b}$$

$$\cdot \qquad \cdot \qquad \cdot$$
$$\cdot \qquad \cdot \qquad \cdot$$
$$\cdot \qquad \cdot \qquad \cdot$$

$$HA^{(n-1)-} \leftrightarrows H^+ + A^{n-} \tag{R2.4n}$$

Note that with the exception of the species H_nA, which can only be an acid, and A^{n-}, which can only be a base, all other species are amphoteric; that is, they are capable of donating or accepting a hydrogen ion.

The problem we must consider in this section has to do with the speciation of H_nA and its associated conjugate bases; in other words, given the series of acid–base reactions represented by Reactions (R2.4a) through (R2.4n), what will be the relative concentrations of H_nA, $H_{n-1}A^-$, ..., A^{n-} in a solution as a function of the solution's pH? We begin addressing this problem by noting that, from our discussion in the previous section on chemical thermodynamics, each of the above equilibria is characterized by an equilibrium constant, K_{an}, such that

$$K_{a1} = \frac{[H^+]\,[H_{n-1}A^-]}{[H_nA]}$$

$$K_{a2} = \frac{[H^+]\,[H_{n-2}A^{2-}]}{[H_{n-1}A^-]}$$

$$\begin{array}{cc} \cdot & \cdot \\ \cdot & \cdot \\ \cdot & \cdot \end{array}$$

$$K_{an} = \frac{[H^+]\,[A^{n-}]}{[HA^{(n-1)-}]} \qquad (2.3.1)$$

It follows from Equation (2.3.1) and continuous substitution that

$$[H_{n-1}A^-] = [H_nA]\left(\frac{K_{a1}}{[H^+]}\right)$$

$$[H_{n-2}A^{2-}] = [H_nA]\left(\frac{K_{a1}K_{a2}}{[H^+]^2}\right)$$

$$\begin{array}{cc} \cdot & \cdot \\ \cdot & \cdot \\ \cdot & \cdot \end{array}$$

$$[A^{n-}] = [H_nA]\left(\frac{K_{a1}K_{a2}\dots K_{an}}{[H^+]^n}\right) \qquad (2.3.2)$$

If we now define C_A as the total concentration of all A-containing species, so that

$$C_A = [H_nA] + [H_{n-1}A^-] + [H_{n-2}A^{2-}] + \cdots + [A^{n-}] \qquad (2.3.3)$$

and we define α_i as the mole fraction of $H_{n-i}A^{i-}$, so that

$$\alpha_i = \frac{[H_{n-i}A^{i-}]}{C_A} \qquad (2.3.4)$$

and

$$\sum \alpha_i = 1 \qquad (2.3.5)$$

we can combine Equations (2.3.2), (2.3.3), and (2.3.4) and obtain expressions for the mole fractions of each of the A-containing species:

$$\alpha_0 = \frac{[H_nA]}{C_A} = \frac{[H^+]^n}{[H^+]^n + K_{a1}[H^+]^{n-1} + K_{a1}K_{a2}[H^+]^{n-2} + \cdots + K_{a1}K_{a2}\cdots K_{an}}$$

$$\alpha_1 = \frac{[H_{n-1}A^-]}{C_A} = \frac{K_{a1}[H^+]^{n-1}}{[H^+]^n + K_{a1}[H^+]^{n-1} + K_{a1}K_{a2}[H^+]^{n-2} + \cdots + K_{a1}K_{a2}\cdots K_{an}}$$

$$\alpha_2 = \frac{[H_{n-2}A^{2-}]}{C_A} = \frac{K_{a1}K_{a2}[H^+]^{n-2}}{[H^+]^n + K_{a1}[H^+]^{n-1} + K_{a1}K_{a2}[H^+]^{n-2} + \cdots + K_{a1}K_{a2}\cdots K_{an}}$$

$$\vdots$$

$$\alpha_n = \frac{[A^{n-}]}{C_A} = \frac{K_{a1}K_{a2}\cdots K_{an}}{[H^+]^n + K_{a1}[H^+]^{n-1} + K_{a1}K_{a2}[H^+]^{n-2} + \cdots + K_{a1}K_{a2}\cdots K_{an}}$$

(2.3.6)

Note that Equation (2.3.6) takes a consistent and regular form. As we progress from one species to the next, the denominator of the right-hand side of the equation remains the same while the numerator changes from the first term in the denominator for the most protonated species (i.e., H_nA), to the second term in the denominator for second most protonated species (i.e., $H_{n-1}A^{n-}$), and so forth, until the last term in the denominator is obtained for the least protonated species (A^{n-}).

The α_i for several acids of interest for biogeochemical cycles are plotted as a function of acidity, pH = $-\log[H^+]$, in Figure 2.1. These plots were generated using Equation (2.3.6) and the thermodynamic data compiled in the Appendix. Note that just as Equation (2.3.6) has a consistent form, the variations of α_i with pH have the same basic characteristics for all polyprotic acids. For each acid, we obtain a progression of bell-shaped curves as the pH is increased from low to high values. At the lowest pH (that is, the highest acidity), the most protonated species is most abundant. As the pH is increased, greater amounts of H^+ are liberated from the acid and the protonated species progressively give way to their conjugate bases. Finally, at the highest pH values, the most dominant species becomes A^{n-}.

An important facet of the acid–base series illustrated in Figures 2.1 has to do with the transition points where any pair of conjugate acid and base are equal in concentration. These transitions occur at very specific and easily predictable points on the graph. If we use the notation that

$$pK_{ai} = -\log(K_{ai})$$ (2.3.7)

(in much the same way as pH is used to represent $-\log([H^+])$, then

$$[H_{n+1-i}A^{1-i}] = [H_{n-1}A^{i-}] \quad \text{and} \quad \alpha_{i-1} = \alpha_i \approx 0.5$$ (2.3.8)

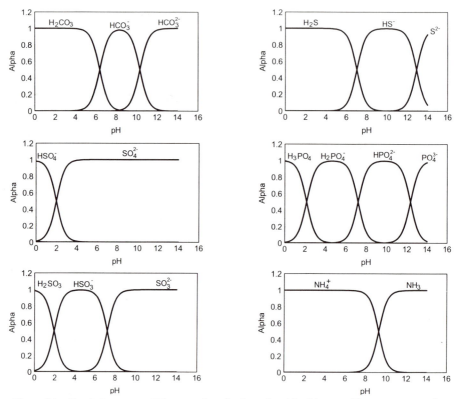

Figure 2.1. Speciation versus pH for a number of polyprotic acids of interest. These curves were derived using Equation (2.3.6) and the data in the Appendix.

when

$$pH = pK_{ai} \qquad (2.3.9)$$

Thus the rudimentary features of the speciation of polyprotic acids and bases as a function of pH can be quickly inferred from a simple knowledge of the values of acid–base dissociation constants of that system.

Because Equation (2.3.6) represents a generic format repeated for all acids, its use affords a quick and easy method for solving otherwise complicated acid–base equilibria. For this reason, many students have found it worthwhile to commit the generalized format for these formulae to memory. For instance, suppose we needed to know the mole fraction of HPO_4^{2-} in a solution containing phosphoric acid (H_3PO_4) and its conjugate bases. An expression for this mole fraction is easily derived from the generalized formula we have derived for a triprotic acid. The denominator in the formula must be given by

$$[H^+]^3 + K_{a1}[H^+]^2 + K_{a1}K_{a2}[H^+] + K_{a1}K_{a2}K_{a3} \qquad (2.3.10)$$

The numerator can be predicted if we recall that the first denominator term from

$$\alpha_i = \frac{[H_{n-i}A^{(n-i)-}]}{C_A}$$

$$\frac{K_{a1}K_{a2}\cdots K_{ai}[H^+]^{n-i}}{[H^+]^n + K_{a1}[H^+]^{n-1} + \cdots + K_{a1}K_{a2}\cdots K_{ai}[H^+]^{n-i} + \cdots + K_{a1}K_{a2}\cdots K_{ai}\cdots K_{an}}$$

Figure 2.2. The mole fraction, α_i, of $H_{n-i}A^{(n-1)-}$, the ith conjugate base of the polyprotic acid H_nA. The mole fraction is easily written down by following a few simple rules. The denominator of the fraction is a sum of $n + 1$ terms: The first term is $[H^+]$ raised to the nth power, the second term is the product of the first acid dissociation constant, K_{a1}, and $[H^+]$ is raised to the $(n-1)$st power, and so on, until the $(n + 1)$st term, which is the product of all the acid dissociation constants. The numerator is simply the term in the denominator that contains $[H^+]$ raised to the $(n - i)$th power.

Equation (2.3.10) is used as the numerator for the most highly protonated species (i.e., H_3PO_4 in this case), and so on. This process leads to the final expression:

$$\alpha_{H_2PO_4^{2-}} = \frac{K_{a1}K_{a2}[H^+]}{[H^+]^3 + K_{a1}[H^+]^2 + K_{a1}K_{a2}[H^+] + K_{a1}K_{a2}K_{a3}} \quad (2.3.11)$$

In general, if we know the general form of the denominator for a polyprotic acid and which term of the denominator to choose for the numerator, we can quickly write down the expression for the mole fraction of any species involved in a stepwise acid dissociation sequence without having to resort to a laborious algebraic derivation (Figure 2.2).

2.4. PHASE TRANSITIONS

Another critical set of equilibria for biogeochemical cycles are those that control transitions between phases. Two types of transitions are of particular interest: transitions between the gas and aqueous phases (i.e., dissolution and evaporation) and between the aqueous and solid phases (i.e., precipitation and solvation). By "transitions between the gas and aqueous phases" we mean reactions of the following sort:

$$X_g \leftrightharpoons X_{aq} \quad (R2.5)$$

where X_g and X_{aq} represent some hypothetical compound X as a gas and as a dissolved species in a liquid water solution, respectively. (By convention reaction pairs between the gas and aqueous phase are written with the gaseous species on the left-hand side and the aqueous-phase species on the right-hand side.) At equilibrium, the relative abundances of X in the two phases is governed by the so-called Henry's law; that is,

$$[X] = K_{H, X} p_X \qquad (2.4.1)$$

which the reader will recognize as just another way of representing Equation (2.2.2) with $K_{H, X}$ being the equilibrium constant. In this case, however, $K_{H, X}$, is usually referred to as the *Henry's law coefficient* and [X] is the solubility of X_g (in moles liter^{-1}) in water at a partial pressure of p_X atm. By requiring equality of units on both sides of Equation (2.4.1), it is easily seen that $K_{H, X}$ must have units of moles liter^{-1} atm^{-1}.

The other phase transition of interest relates to the dissolution of a solid compound into an aqueous solution. Most often this occurs as a result of a reaction that liberates two dissolved ionic species from a single insoluble precipitate, that is,

$$AB_s \leftrightharpoons A^+ + B^- \qquad (R.2.6)$$

where A^+ and B^- represent ions in solution and AB a solid precipitate. Because solids have unit activity, the equilibrium relationship is governed by

$$K_{S, AB} = [A^+][B^-] \qquad (2.4.2)$$

and the equilibrium constant, $K_{S, AB}$, is usually referred to as the *solubility product* for compound AB and has units of (moles liter^{-1})^2.

2.5. APPLICATION OF EQUILIBRIA TO THE CO$_2$–H$_2$O–CALCIUM SYSTEM

The chemical system involving atmospheric carbon dioxide (CO_2), water (H_2O), and calcium (Ca) plays a central role in biogeochemical cycles. As we know, C is the primary building block of the biosphere and CO_2 is required by all green plants to carry out photosynthesis. In the ocean, the dissolution of atmospheric CO_2 and the resulting acid–base chemistry involving carbonic acid (H_2CO_3) and its conjugate bases (HCO_3^- and CO_3^{2-}) can affect another important biospheric process, namely, the formation of solid $CaCO_3$ by shell-forming organisms. Not only does the formation of these $CaCO_3$ shells represent a significant source of atmospheric CO_2, but it is also the primary source for one of the major rock-forming minerals found in marine sediments and on land. For these reasons, an understanding of these interactions is central to our being able to quantitatively understand biogeochemical cycles. In this section we explore how the thermodynamic relationships we derived in the previous sections can be used to characterize the speciation and interactions of this system. We will do this by considering a series of specific problems and then applying the appropriate equations and thermodynamic constants (obtained from the Appendix) to their solution.

2.5.1. Pure Carbonate System

In our first example, we consider a system involving only H_2O and CO_2. Strictly speaking, description of this system requires consideration of five equilibrium reactions: the acid dissociation of H_2O,

$$H_2O \leftrightharpoons H^+ + OH^- \tag{R2.2}$$

the dissolution of gaseous carbon dioxide, $(CO_2)_g$, in water to form dissolved carbon dioxide, $(CO_2)_{aq}$,

$$(CO_2)_g \leftrightharpoons (CO_2)_{aq} \tag{R2.7}$$

the hydration of $(CO_2)_{aq}$ to form carbonic acid,

$$(CO_2)_{aq} \leftrightharpoons H_2CO_3 \tag{R2.8}$$

and the two acid dissociation reactions,

$$H_2CO_3 \leftrightharpoons HCO_3^- + H^+ \tag{R2.9}$$

$$HCO_3^- \leftrightharpoons CO_3^{2-} + H^+ \tag{R2.10}$$

However, this system is typically simplified into four reactions by defining a hypothetical species $H_2CO_3^*$ as the total amount of undissociated carbon dioxide in solution, such that

$$[H_2CO_3^*] = [CO_2]_{aq} + [H_2CO_3] \tag{2.5.1}$$

Reactions (R2.7) and (R2.8) are then combined into a single hypothetical dissolution reaction,

$$(CO_2)_g \leftrightharpoons H_2CO_3^* \tag{R2.7'}$$

and (R2.9) is rewritten in terms of $H_2CO_3^*$,

$$H_2CO_3^* \leftrightharpoons HCO_3^- + H^+ \tag{R2.9'}$$

The Problem: Given the above four-reaction set (R2.2, R2.7', R2.9', and R2.10) and a constant partial pressure for atmospheric CO_2 gas of 355 ppmv (i.e., 355×10^{-6} atm), determine C_C, the total concentration of dissolved carbon species in equilibrium with gaseous CO_2.

The Solution: First note that, given a $(CO_2)_g$ partial pressure, $p(CO_2)$, of 355 ppmv, we can easily express the concentration of $H_2CO_3^*$ in terms of this partial pressure by applying Henry's law [Equation (2.4.1)] to Reaction (R2.7'):

$$[H_2CO_3^*] = K_{H, CO_2} \, p(CO_2) \tag{2.5.2}$$

We can then apply Equation (2.3.6) for a diprotic acid (i.e., $n = 2$) to obtain

$$[H_2CO_3^*] = C_C \frac{[H^+]^2}{[H^+]^2 + K_{H2CO3,1} [H^+] + K_{H2CO3,1}K_{H2CO3,2}} \tag{2.5.3}$$

and combining Equations (2.5.2) and (2.5.3), we obtain an expression for C_C:

$$C_C = K_{H,CO_2} \, p(CO_2) \frac{[H^+]^2 + K_1[H^+] + K_1K_2}{[H^+]^2} \tag{2.5.4}$$

There is, however, a problem; while Equation (2.5.4) defines the total concentration of dissolved carbon, it does so in terms of the hydrogen ion concentration.

To actually determine C_C, we must also determine $[H^+]$. To calculate $[H^+]$, we need to include two additional relationships. The first relationship is a statement of charge conservation—that is, that the sum of all the positive ions in solution must equal the sum of all the negative ions. For our simple problem, that equation becomes

$$[H^+] = [HCO_3^-] + 2[CO_3^{2-}] + [OH^-] \tag{2.5.5}$$

The second relationship expresses $[OH^-]$ in terms of $[H^+]$ using the equilibrium constant for (R2.2), that is,

$$[OH^-] = \frac{K_{H2O}}{[H^+]} \tag{2.5.6}$$

Equation (2.5.5) can now be combined with (2.3.2) for a diprotic acid and (2.5.6) to obtain

$$[H^+] = K_{H,CO2}\, p(CO_2) \left(\frac{K_{H2CO3,1}}{[H^+]} \right)$$
$$+ 2\, K_{H,CO2}\, p(CO_2) \left(\frac{K_{H2CO3,1} K_{H2CO3,2}}{[H^+]^2} \right) + \frac{K_{H2O}}{[H^+]} \tag{2.5.7}$$

Equation (2.5.7) can be rearranged to yield a third-order polynomial in $[H^+]$ that can be solved using the standard equation for a third-order polynomial and a hand-held calculator or a computer spreadsheet. However, a very good approximation to the solution can be obtained through a simplification of Equation (2.5.7). It turns out, for the conditions of our problem, that

$$[HCO_3^-] \gg 2[CO_3^{2-}]$$
$$[HCO_3^-] \gg [OH^-] \tag{2.5.8}$$

We can therefore drop the terms arising from CO_3^{2-} and OH^- in the charge balance equation, and we obtain

$$[H^+] \approx [HCO_3^-] = K_{H,\,CO2}\, p(CO_2)\, \frac{K_{H2CO3,1}}{[H^+]} \tag{2.5.9}$$

and thus

$$[H^+] = \sqrt{K_{H,\,CO2}\, p(CO_2)\, K_{H2CO3,1}} \tag{2.5.10}$$

Substituting for $K_{H,CO2}$ and $K_{H2CO3,1}$ from the Appendix and setting $p(CO_2) = 3.55 \times 10^{-6}$, we obtain

$$[H^+] = 2.38 \times 10^{-6} \text{ mole liter}^{-1}$$
$$pH = 5.62 \tag{2.5.11}$$

Substituting this hydrogen ion concentration into Equation (2.5.4), we are finally able to obtain the solution to our problem:

$$C_C = 2.38 \times 10^{-6} \text{ mole liter}^{-1} \tag{2.5.12}$$

Note that C_C is calculated to be essentially equal to $[H^+]$. Moreover, it is easily shown from Equations (2.3.2) and (2.5.6) that

$$[HCO_3^-] = 2.38 \times 10^{-6} \text{ mole liter}^{-1}$$
$$\approx C_C >> [OH^-]$$
$$>> 2[CO_3^{2-}] \qquad (2.5.13)$$

thereby confirming the validity of our assumption that the concentrations of OH^- and $CO_3^=$ are small and can be neglected (see Text Box 2.3).

2.5.2. The Calcium Carbonate System

For our second illustrative example, we add calcium to the system. The addition of calcium requires that we consider an additional reaction, the phase transition between solid calcium carbonate and aqueous calcium and carbonate ions:

$$(CaCO_3)_s \leftrightharpoons Ca^{2+} + CO_3^{2-} \qquad (R2.11)$$

The Problem: Calculate C_C for a solution in equilibrium with an atmosphere containing 355 ppmv of CO_2 and a solid precipitate of $CaCO_3$.

The Solution: This problem is quite similar to the previous one, except that, with the addition of (R2.11), we now have an additional unknown variable to solve for, namely, $[Ca^{2+}]$. The presence of an additional unknown in our system requires that we add an additional equation. This additional equation is simply the applica-

TEXT BOX 2.3	ACID RAIN AND THE CARBON DIOXIDE–CARBONIC ACID SYSTEM

In Section 2.5.1 we found that a water solution in equilibrium with 355 ppmv of carbon dioxide will have a pH of 5.6. It turns out that this pH is of some geochemical significance. Because our atmosphere contains about 355 ppmv of gaseous CO_2, liquid water solutions in equilibrium with the atmosphere should generally all attain a pH of 5.6 if no other acids or bases are present. For this reason, cloud water and rainwater with a pH of 5.6 are usually referred to as *neutral precipitation*, even though pure water has a pH of 7. By the same token, *acid rain* refers to rainwater with a pH below 5.6 (see Figure 2.3). Quite often the extra acidity in acid rain is sulfuric and/or nitric acid arising from S- and N-containing air pollutants. Because of concern that acid rain may be harmful to sensitive ecosystems in the northeastern United States, the Clean Air Act Amendments of 1990 prescribed reductions in the emissions of sulfur oxides and nitrogen oxides from power plants; implicit in these regulations is the belief that reductions in these emissions will bring the acidity of rainfall over the United States closer to its "neutral" value of 5.6 and will be less harmful to terrestrial ecosystems.

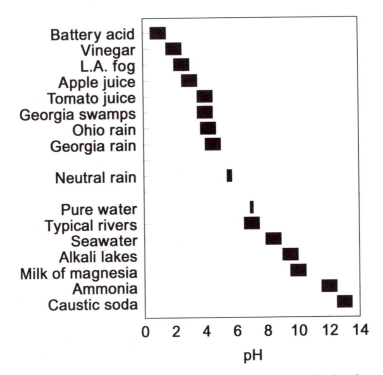

Figure 2.3. The pH scale for various commonly encountered solutions including that of *pure water* (7.0), *neutral rain* (5.6), and typical *acid rain* from Georgia and Ohio and *acid fog* from Los Angeles.

tion of Equation (2.4.2) to the dissolution of $CaCO_3$; that is,

$$K_{S,CaCO_3} = [C_a^{2+}] [CO_3^{2-}] \qquad (2.5.14)$$

Moreover, because of the presence of calcium ions, we must revise our charge balance equation:

$$[H^+] + 2 [Ca^{2+}] = [HCO_3^-] + 2 [CO_3^{2-}] + [OH^-] \qquad (2.5.15)$$

Combining Equation (2.5.14) with (2.3.6) for a diprotic acid, it is found that

$$[Ca^{2+}] = \frac{K_{S,CaCO_3}}{[CO_3^{2-}]}$$

$$= K_{S, CaCO_3} \left(\frac{[H^+]^2}{K_{H, CO_2}\, p(CO_2)\, K_{H2CO3,1}\, K_{H2CO3,2}} \right) \qquad (2.5.16)$$

Substituting this expression and the expressions for $[HCO_3^-]$, $[CO_3^{2-}]$, and $[OH^-]$ derived previously into charge balance Equation (2.5.15), we obtain

$$[H^+] + 2 \frac{K_{S,CaCO_3}[H^+]^2}{K_{H,CO_2} p(CO_2) K_{H_2CO_3,1} K_{H_2CO_3,2}} = \frac{K_{H,CO_2} p(CO_2) K_{H_2CO_3,1}}{[H^+]}$$

$$+ 2 \frac{K_{H,CO_2} p(CO_2) K_{H_2CO_3,1} K_{H_2CO_3,2}}{[H^+]^2} + \frac{K_{H_2O}}{[H^+]} \qquad (2.5.17)$$

Equation (2.5.17) yields a fourth-order polynomial in $[H^+]$ that can be solved by successive approximation using a standard spreadsheet or a hand-held calculator. However, as in the previous example, the problem can be simplified considerably by making the appropriate approximations. In this case it turns out that

$$[HCO_3^-] >> [CO_3^{2-}]$$

$$[HCO_3^-] >> [OH^-] \qquad (2.5.18)$$

$$2[Ca^{2+}] >> [H^+]$$

Thus, by dropping the $[CO_3^{2-}]$, $[OH^-]$, and $[H^+]$ terms from Equation (2.5.15), we obtain an approximate charge balance equation of

$$2 \frac{K_{S,CaCO_3}[H^+]^2}{K_{H,CO_2} p(CO_2) K_{H_2CO_3,1} K_{H_2CO_3,2}} \approx \frac{K_{H,CO_2} p(CO_2) K_{H_2CO_3,1}}{[H^+]} \qquad (2.5.19)$$

which is easily solved to yield

$$[H^+] = \sqrt[1/3]{\frac{(K_{H,CO_2} p(CO_2) K_{H_2CO_3,1})^2 K_{H_2CO_3,2}}{2 K_{S,CaCO_3}}} \qquad (2.5.20)$$

Using the appropriate equilibrium constants from the Appendix and setting $p(CO_2) = 355 \times 10^{-6}$ atm, we obtain

$$[H^+] = 4.2 \times 10^{-9} \text{ mole liter}^{-1}$$

$$pH = 8.4 \qquad (2.5.21)$$

Substitution of this result into Equation (2.5.4) yields the solution to our problem:

$$C_C = \left(\frac{2 K_{H,CO_2} K_{H_2CO_3,1} K_{S,CaCO_3} p(CO_2)}{K_{H_2CO_3,2}} \right)^{1/3}$$

$$= 1.35 \times 10^{-3} \text{ mole liter}^{-1} \qquad (2.5.22)$$

Similar to the example in Section 2.5.1, it turns out that the major carbon-containing species is HCO_3^-; however, in this case $[HCO_3^-] \cong [C_C] \cong 0.5 [Ca^{2+}] >> [H^+]$ (see Text Box 2.4).

2.6. REDOX CHEMISTRY

When an atom in any chemical species gains or loses electrons, it is said to undergo a change in its *oxidation state*. An atom's oxidation state is defined as the difference between the number of nuclear protons and electrons associated with that atom.

When an atom has the same number of protons and electrons, its oxidation state, by definition, is zero (0). As this atom gains electrons, its oxidation state decreases from 0 to -1, and then to -2, and so on. Because the gain of electrons causes a more negative or reduced oxidation state, this process is referred to as *reduction*. Conversely, as an atom loses electrons, its oxidation state increases from 0 to $+1$, and then to $+2$, and so on. Because atoms almost always lose electrons when they react with oxygen, the loss of electrons is referred to as *oxidation*. In this section we review the essential features of reduction and oxidation, or simply *redox equilibria*.

TEXT BOX 2.4   **THE ROLE OF THE CARBONATE SYSTEM IN BUFFERING OCEAN PH**

In Section 2.5.2 we found that a solution in equilibrium with solid $CaCO_3$ and atmospheric CO_2 (at 355 ppmv) will attain a pH of about 8.4. Interestingly, 8.4 is also the pH typically encountered in surface seawater. At first, you might not find this too surprising, because in our example the ocean is close to being saturated in $CaCO_3$ and it is also essentially in equilibrium with an atmosphere containing about 355 ppmv of CO_2. Thus, on the basis of the solution in Section 2.5.2, one might expect the ocean pH to be about 8.4.

But recall that our example considered a very simple solution—that is, one in equilibrium with $CaCO_3$ and atmospheric CO_2 only, with no other acids or bases. Since the ocean also has a pH of 8.4, can we conclude that it, similar to the simple solution considered in our example, is devoid of other acidic or basic species? The answer is no. But the ocean still has a pH of about 8.4 because the calcium carbonate–atmospheric CO_2 system provides a very effective buffer for the ocean that stabilizes its pH. As illustrated in Figure 2.4, as long as the amount of strong acid (HA) or base (BOH) does not exceed the millimolar level, the pH of a solution in equilibrium with solid $CaCO_3$ and 355 ppmv of atmospheric CO_2 remains essentially constant at 8.4. For instance, as a strong acid is added to the solution, its dissociation adds an H^+ ion to the solution:

$$HA \leftrightarrows H^+ + A^-$$

But rather than stoichiometrically lowering the pH, the extra H^+ ion is, to a first approximation, removed via the dissolution of a molecule of $CaCO_3$ and the formation of a HCO_3^- ion:

$$(CaCO_3)_s \leftrightarrows Ca^{2+} + CO_3^{2-}$$
$$CO_3^{2-} + H^+ \leftrightarrows HCO_3^-$$

The result: very little net change in pH. Similarly, if a base is added, a molecule of $CaCO_3$ is removed from solution, and in the process an additional H^+ ion is added. This H^+ ion neutralizes the OH^- ion added from the dissociation of BOH, and the result is a nearly unchanged pH. When the amount of strong acid or base exceeds the millimolar level, the solution remains buffered but not as effectively and so the pH begins to diverge from 8.4. Do you know why? How many molecules of $CaCO_3$ must be dissolved (precipitated) for each molecule of HA (BOH) added in this case?

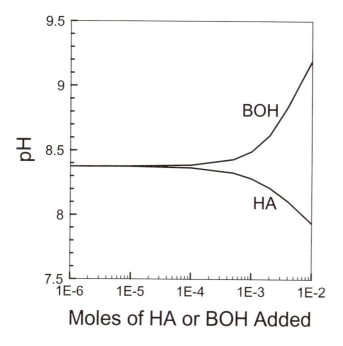

Figure 2.4. The buffering effect of the $CaCO_3$–CO_2 system. The pH of a 1-liter solution in equilibrium with solid $CaCO_3$ and 355 ppmv of gaseous CO_2 is plotted as a function of the number of moles of a strong acid (HA) or a strong base (BOH) added to the solution. You can examine this buffering effect yourself by using a modified form of the charge balance Equation (2.5.15) in which $[B^+]$ or $[A^-]$ are added to the left-hand or right-hand side of the equation, respectively. By specifying various values for $[B^+]$ or $[A^-]$ and solving for $[H^+]$ using any appropriate method, you should be able to duplicate the results given here.

2.6.1. Oxidation States

When a chemical bond is formed between two atoms, there is generally an equal and opposite change in the oxidation states of the two atoms. The more *electronegative* atom (i.e., the atom with the greater affinity for electrons) will generally experience a decrease in its oxidation state, while the less electronegative atom will increase its oxidation state. This shift in oxidation state can occur via an actual transfer of electrons to form an *ionic bond* when the difference in the electronegativities of the two atoms is sufficiently large, or via an unequal sharing of electrons in a *polar covalent bond* when the difference in the electronegativities of the two atoms is smaller. For example, consider the reaction of calcium (Ca) with oxygen (O_2) via (R2.12):

$$Ca + \tfrac{1}{2}O_2 \rightarrow CaO \tag{R2.12}$$

In this reaction, a Ca atom donates two electrons ($2e^-$) to an oxygen atom, resulting in the formation of calcium oxide (CaO), a solid consisting of a Ca^{2+} ion (with an oxidation state of $+2$) and an O^{2-} ion (with an oxidation state of -2) held together by an ionic bond. Alternatively, consider the formation of liquid water (H_2O) from hydrogen (H_2) and (O_2) via (R2.13):

$$H_2 + \tfrac{1}{2}O_2 \rightarrow H_2O \qquad \text{(R2.13)}$$

In this reaction, electrons are not actually donated from one atom to another. Instead, an electron from each hydrogen atom is shared with the oxygen atom. The result is an H–O–H molecule held together by a covalent bond between each of the hydrogen atoms and the oxygen atom. Although these covalent bonds do not involve an actual electron donation, the electrons involved in each of the bonds are not shared equally either. Because oxygen is more electronegative than hydrogen, there is an increased electron density around the O atom in H–O–H and a decreased electron density around the H atoms, relative to the respective atoms in their elemental states. For this reason, the O atom in water is considered to have an oxidation state of -2 and the H atoms are considered to have oxidation states of $+1$, even though they do not actually exist as ions.

Even though the oxidation state of an atom in a compound containing ionic bonds is fairly evident, the oxidation states of atoms in compounds with covalent bonds are not. However, chemists have developed a set of "rules" that can be used as a guide for determining oxidation states in these cases:

1. The sum of the oxidation states of all the atoms in a chemical species must equal the net charge of that species.
2. Atoms in a pure element (e.g., O_2, N_2) have an oxidation state of 0.
3. Oxygen is nearly always assigned an oxidation number of -2 (except when it appears as a pure oxygen allotrope or as a peroxide).
4. The Group 1A and 2A elements (the metals) tend to lose 1 and 2 electrons in their reactions, and thus attain oxidation states of $+1$ and $+2$, respectively.
5. Halogens (Group 7A elements) almost always have -1 oxidation states, unless bonded to O.
6. Hydrogen can gain or lose one electron in compound formation. Generally, hydrogen will have a $+1$ oxidation state when bonded to nonmetals but will have -1 oxidation state when bonded to metals.

The oxidation states of C, H, O, N, S, and P in some of the principal compounds we will consider here are listed in Table 2.2. Using the rules listed above, you should be able to quickly produce a similar table on your own, with perhaps one exception—the oxidation state of oxygen in H_2O_2. Can you rationalize an oxidation state of -1 for oxygen in this compound in light of rules 2 and 3 listed above? (see Text Box 2.5).

2.6.2. Redox Half-Reactions

The tendency for an atom or group of atoms to accept electrons (i.e., its electron affinity) is an important thermochemical property of that species that ultimately determines its ability to react with other chemical species and form chemical bonds. This property is quantitatively expressed in a variety of ways (see Text Box 2.6); we will use the *redox half-reaction* representation. A redox half-reaction describes the reduction of an oxidized species (OX_i) by "n" electrons (ne^-) to form the corresponding reduced compound (RED_i):

$$OX_i + ne^- \leftrightarrows RED_i \qquad (R.2.14)$$

From our previous discussion of chemical equilibria, we know that the equilibrium constant for this reaction must be given by

TABLE 2.2
Oxidation States of C, H, O, N, S, and P in Various Compounds of Biogeochemical Interest

Compound	Element					
	C	H	O	N	S	P
CH_4	-4	$+1$				
CH_3OH	-2	$+1$	-2			
CO	$+2$		-2			
CO_2, H_2CO_3, HCO_3^-, CO_3^{2-}	$+4$	$+1$	-2			
H_2O		$+1$	-2			
H_2O_2		$+1$	-1			
O_2			0			
NH_3, NH_4^+		$+1$		-3		
NH_2OH		$+1$	-2	-1		
N_2				0		
N_2O			-2	$+1$		
NO			-2	$+2$		
HONO		$+1$	-2	$+3$		
NO_2			-2	$+4$		
N_2O_5, HNO_3, NO_3^-		$+1$	-2	$+5$		
CH_3SCH_3	-2	$+1$			-2	
OCS, H_2S, HS^-, S^{2-}	$+4$	$+1$	-2		-2	
FeS_2					-1	
S_8					0	
SO_2, H_2SO_3, HSO_3^-, SO_3^{2-}		$+1$	-2		$+4$	
SO_3, H_2SO_4, HSO_4^-, SO_4^{2-}		$+1$	-2		$+6$	
PH_3		$+1$				-3
H_3PO_3, $H_2PO_3^-$		$+1$	-2			$+3$
H_3PO_4, $H_2PO_4^-$, HPO_4^{2-}, PO_4^{3-}		$+1$	-2			$+5$

| TEXT BOX 2.5 | REDOX REACTIONS |

Having established the concepts of oxidation state and reduction and oxidation, it is interesting to return the acid–base and phase change reactions we considered earlier in this chapter and consider the oxidation states of the reactants and products in these reactions. If you do this, you should be able to discover an interesting characteristic of these reactions: They are completely devoid of any change in the oxidation states of the elements involved. For instance, consider the equilibria between gaseous CO_2 and aqueous-phase carbonic acid (H_2CO_3) and its base conjugates, bicarbonate ion (HCO_3^-) and carbonate ion (CO_3^{2-}), involving (R2.2.4), (R2.2.5), and (R.2.2.6). From Table 2.2 we see that, for each of these species, the oxidation state for carbon and oxygen are always +4 and −2, respectively. Clearly, redox chemistry—the chemistry of reduction and oxidation—represents a distinctly different kind of chemistry from that involving acid–base and phase change equilibria. In fact, as we will see later, redox chemistry comprises the group of reactions that ultimately make possible the storage, transfer, and utilization of chemical energy and, as a result, plays a critical role in the biogeochemical cycles we will study here.

$$K_{e,OX_i} = \frac{\{RED_i\}}{\{OX_i\}\{e^-\}^n} \tag{2.6.1}$$

As the electron affinity of OX_i increases, the more the equilibrium of (R.2.14) is shifted to the right and the larger the K_{e,OX_i} becomes. Thus redox half-reaction equilibrium constants provide a quantitative method of comparing the electron affinities of the various elements, and compounds of these elements form; a listing of these constants for the principal elements and compounds of interest to us here can be found in the Appendix.

The formalism adopted by chemists for redox half-reactions is quite similar to that adopted for acid–base reactions described in Section 2.3 [see Table 2.3]. Just as we defined acids as protons donors and bases as proton acceptors, we can think of a *reducing agent* as an electron donor and think of an *oxidizing agent* as a proton acceptor. So for instance in (R2.14), RED_i is a reducing agent and OX_i is an oxidizing agent.

The analogy with acid–base chemistry is carried further by treating electron activity in aqueous solutions in a manner similar to that of H^+ activity. Thus, just as we define pH as being equal to the negative log of H^+ activity (i.e., pH = $-\log\{H^+\}$), we can define pe as follows:

$$pe = -\log\{e^-\} \tag{2.6.2}$$

Moreover, just as pH is a measure of the acidity of a solution, pe can be thought of as a measure of the oxidizing potential of the solution. Recall from Section 2.3 that the larger the pH, the less acidic the solution and the greater the tendency for species in that solution to exist in their conjugate base forms. Similarly, the larger the pe,

TEXT BOX 2.6	RELATIONSHIP BETWEEN ELECTROCHEMICAL POTENTIAL AND HALF-REACTION EQUILIBRIUM CONSTANT

The affinity of an oxidized species for electrons can be expressed using two different but equivalent formulations: one involving the electrochemical potential and the other using the redox half-reaction described in the text. The electrochemical potential formulation is based on the Nernst equation:

$$E_h = E^0 - \left(2.3\,\frac{RT}{nF}\right)\log\left[\frac{\{RED_i\}}{\{OX_i\}}\right] \quad (2.6.4)$$

where E_h is the electrochemical potential (in volts) required to maintain the activities of RED_i and OX_i at the given levels for a temperature T, E^0 is the electrochemical potential (in volts) for standard conditions (i.e., $\{RED\}_i = \{OX_i\} = 1$), R is the gas constant (=8.314 J mole^{-1} K^{-1}), n is the number of electrons needed to convert OX_i to RED_i, and F is the Faraday constant (96,487 coulomb mole^{-1}).

Alternatively, the same quantitative information can be expressed using the equilibrium constant for the appropriate redox half-reaction; namely, (R2.5.3):

$$K_{e,OXi} = \frac{\{RED_i\}}{\{OX_i\}\{e^-\}^n} \quad (2.6.5)$$

If we now take the log of equation (2.6.5) and rearrange terms, we obtain

$$\log\left[\frac{\{RED_i\}}{\{OX_i\}}\right] = \log K + n\log\{e^-\} = \log K - n\cdot pe \quad (2.6.6)$$

where $pe = -\log\{e^-\}$. But from Equation (2.6.4) we obtain

$$\log\left[\frac{\{RED_i\}}{\{OX_i\}}\right] = E^0(nF/2.3RT) - E_h(nF/2.3RT) \quad (2.6.7)$$

Equating Equations (2.6.6) and (2.6.7), we find that

$$pe = E_h\left(\frac{F}{RT}\right) \quad (2.6.8)$$

and

$$\log K = E^0\left(\frac{nF}{2.3RT}\right) \quad (2.6.9)$$

Most likely as you move from one reference book to another, you will encounter instances where redox equilibria are described using a half-reaction equilibrium constant and pe as the master variable and other instances where E^0 and E_h are used. This should not present a problem as long as you remain consistent in your own work and remember to use the appropriate conversion equations to go from one notation to the other.

TABLE 2.3
Analogies Between Acid–Base and Redox Equilibria

	Type of Equilibria	
	Acid–Base	Redox
Generic reaction[a]:	$HA \rightleftarrows H^+ + A^-$	$OX_i + ne^- \rightleftarrows RED_i$
Species:	HA $\quad$ A^- Acid $\quad$ Base (proton donor) $\quad$ (proton acceptor)	OX_i $\quad$ RED_i Oxidizer $\quad$ Reducer (electron $\quad$ (electron donor) acceptor)
Master variable:	$pH = -\log\{H^+\}$ (measure of acidity) High pH indicates low acidity and tendency for species to exist in their conjugate base forms.	$pe = -\log\{e^-\}$ (measure of oxidation) High pe indicates high level of oxidation and tendency for species to exist in oxidized form.

[a]Note the difference in the formalism used to write acid–base and redox equilibria. Acid–base reactions are written as acid *dissociation* reactions, while redox half-reactions are written as electron *association* reactions.

the lower the electron activity in the solution and, from Equation (2.6.1), the more likely it is for species to exist in their oxidized forms.

There is, however, one important difference between the formalism for acid–base chemistry and redox half-reactions. While acid–base equilibria are typically written as acid dissociation reactions with H^+ appearing on the right-hand side [e.g., (R2.3)], redox half-reactions are written as association or stability reactions with e^- appearing on the left-hand side. Thus, strictly speaking, a redox half-reaction is most closely analogous to the reaction between a base and H^+ to produce an acid.

2.6.3. Redox Equilibria

One question that has perhaps risen in the minds of some readers has to do with why the term "half-reaction" is used for reactions such as (R2.14). They are referred to as half-reactions because these reactions actually represent only half of the total redox reaction that usually occurs in natural systems. Although the redox half-reaction provides a convenient formalism for describing electron affinities and oxidation states, half-reactions do not occur in natural systems because electrons do not actually exist as such in these systems. Instead, changes in oxidation states in real systems require the simultaneous reduction of one compound and oxidation of another; that is,

$$OX_1 + RED_2 \rightleftarrows RED_1 + OX_2 \qquad (R2.15)$$

When (R2.15) proceeds from the left-hand side to the right-hand side, OX_1 gains one or more electrons and thus acts as the oxidizing agent, while RED_2 loses one or more electrons and thus acts as the reducing agent. Of course, when (R2.15) proceeds from the right to the left, the opposite applies.

The reader will note that (R2.15) is nothing more than the sum of two redox half-reactions,

$$OX_1 + ne^- \rightleftarrows RED_1 \qquad (R2.16)$$

and

$$OX_2 + ne^- \leftrightharpoons RED_2 \qquad (R2.17)$$

with (R2.16) proceeding from left to right and (R2.17) proceeding from right to left. For this reason, reactions such as (R2.15) are referred to as redox reactions, and (R2.16) and (R2.17) are referred to as redox half-reactions. Finally note that from Equations (2.2.1) and (2.6.1), it follows that the equilibrium constant for any redox reaction is simply given by the ratio of the equilibrium constants for the appropriate redox half-reactions (n must be the same in both half-reactions). For instance, the equilibrium constant for (R2.15) is given by

$$K = \frac{[OX_2][RED_1]}{[RED_2][OX_1]} = \frac{K_{e,OX_1}}{K_{e,OX_2}} \qquad (2.6.3)$$

where K_{e,OX_1} and K_{e,OX_2} are the equilibrium constants for redox half-reactions (R2.16) and (R2.17), respectively. From Equation (2.6.3) and the fact that the equilibrium constants for redox half-reactions are a measure of the electron affinities of the relevant species, we see that the degree to which the equilibrium for redox reactions is shifted to one side or the other is determined by the relative electron affinities of the oxidized species involved in the reaction. For instance, referring back to (R2.14), we see that for a highly electronegative OX_1 and a less electronegative OX_2, we would expect the ratio of the equilibrium constants for the two half-reactions to be relatively large and, thus, the equilibrium for the reaction to be shifted to the right.

2.7. THE pe–pH STABILITY DIAGRAMS

In the preceding sections, we have seen how pH and pe act to determine the chemical forms that elements will tend to take, with pH largely controlling the degree of protonation or acidity of the species and pe controlling the oxidation state. In fact, we can think of pH and pe as comprising a two-dimensional *phase space* that determines the acidity and oxidation state of elements under equilibrium conditions. As pH and pe vary within an environment, we effectively move from one regime of this phase space to another, and the oxidation state and acidity of the chemical species within that environment vary accordingly (see Figure 2.5). For instance, consider the element sulfur (S). Referring back to Table 2.2, we see that the oxidation state of this element varies from its most reduced state of -2 to its most oxidized state of $+6$. Moreover, in its -2 oxidation state the element's chemical makeup can vary from its most acidic form of hydrogen sulfide (H_2S) to its most deprotonated or basic form as the sulfide ion (S^{2-}), while in its $+6$ oxidation state it can vary from its most acidic form of sulfuric acid (H_2SO_4) to its most basic form as the sulfate ion (SO_4^{2-}). Thus as we move in a clockwise direction around the four corners of the pe-pH two-dimensional phase space, we would expect the predominant S species at equilibrium to vary from

1. H_2SO_4 in the upper left-hand corner, where pH is lowest and pe highest, to
2. SO_4^{2-} in the upper right-hand corner, where both pH and pe are highest, to

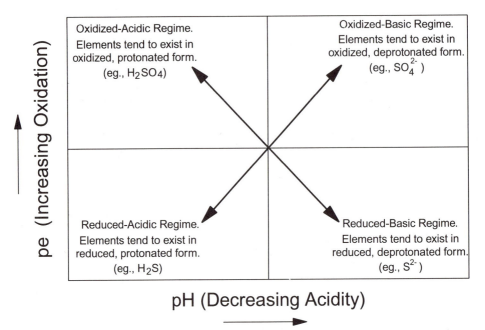

Figure 2.5. The pe–pH stability diagram. Because of chemical thermodynamics, pH and pe can be thought of master variables that determine the equilibrium speciation of any element with regard to its acidity and oxidation state. These relationships can be graphically depicted in a pe–pH stability diagram where pH and pe comprise the x and y coordinates and lines drawn within this diagram describe boundaries between regions of this phase space where specific chemical forms of the element are thermodynamically favored over other forms of the element.

3. S^{2-} in the lower right-hand corner, where pH is highest and pe are lowest, to

4. H_2S in the lower left-hand corner, where both pH and pe are lowest.

The qualitative depiction in Figure 2.5 of the general trends in the equilibrium speciation of the elements as one moves within this two-dimensional pe–pH phase space is given a more quantitative representation in a pe–pH stability diagram. In this diagram, acid–base and redox half-reaction equilibria are used to determine the most abundant chemical form of an element at each point in the pe–pH phase space. Lines are drawn in the two-dimensional phase space, to identify transition boundaries between two chemical forms of an element. By standard convention, the boundaries or lines are generally drawn along the locus of pe–pH values where the activities of the two chemical species that predominate on either side of the boundary are equal if the species are in the same phase. (Otherwise, activities are arbitrarily assigned to determine the exact location of the boundary.) We will find in our study of biogeochemical cycles that these pe–pH stability diagrams provide a useful and easily assimilated picture of how the speciation of any element will vary as environmental conditions of pH and pe vary. For this reason it is important that the reader have a basic understanding of how these diagrams are constructed and what information they convey. In the subsections below, we illustrate the construction and interpretation of such diagrams.

2.7.1 pe–pH Stability Diagram For Water

The stability diagram for water (H_2O) is obtained by considering the redox equilibrium between water and molecular oxygen (O_2) via (R2.18)

$$\tfrac{1}{4}O_2 + H^+ + e^- \leftrightarrows \tfrac{1}{2}H_2O \qquad (\log K_{2.18} = 20.75) \qquad \text{(R2.18)}$$

and the equilibrium between a hydrogen ion and molecular H_2 via (R2.19)

$$H^+ + e^- \leftrightarrows \tfrac{1}{2}H_2 \qquad (\log K_{2.19} = 0) \qquad \text{(R2.19)}$$

(The equilibrium constants indicated to the right of each of the half-reactions are taken from the Appendix.)

From (R2.18), we know that

$$K_{2.18} = 10^{20.75} = \frac{1}{p_{O_2}{}^{0.25}\,[H^+]\,[e^-]} \qquad (2.7.1)$$

or, taking the log of both sides,

$$\log K_{2.18} = 20.75 = -\tfrac{1}{4}\log p_{O_2} + pH + pe \qquad (2.7.2)$$

When $p_{O2} = 1$ atm, Equation (2.7.2) reduces to

$$pe = 20.75 - pH \qquad (2.7.3)$$

Similarly, for (R2.19) it easily shown that

$$\log K_{2.19} = 0 = \tfrac{1}{2}\log p_{H_2} + pH + pe \qquad (2.7.4)$$

and when $p_{H2} = 1$ atm

$$pe = 0 - pH \qquad (2.7.5)$$

Because water is assigned an activity of 1 (see Text Box 2.1), Equations (2.7.3) and (2.7.5) describe the values of pH and pe that yield equal activities of O_2 and H_2O and of H_2 and H_2O, respectively. Following the convention noted above, we can construct our pe–pH stability diagram for water by plotting these two equations in pe–pH phase space as illustrated in Figure 2.6. Between the two lines, H_2O is stable with respect to 1 atm of gaseous O_2 and H_2. However, above the upper line, H_2O will spontaneously decompose into O_2, and below the lower line, H_2O will decompose into H_2. Because the two lines in Figure 2.6 represent the boundaries of the stability field of water, they also represent the pe–pH boundaries within which all of aqueous geochemistry must occur. It should be noted in this regard that the diagram is essentially plotted on a log–log scale since every unit variation in pH and pe represents a factor of 10 variation in H^+ and e^- activity. Thus the stability field of water affords a wide range of potential aquatic environments for geochemical processes to occur. In fact, as can be seen in Figure 2.7, where the pH and pe values typically encountered in various aquatic environments are indicated, natural water systems cover a significant fraction of this total stability field.

2.7.2 A pe–pH Stability Diagram for a Simplified Sulfur System

For our second example, we construct a stability diagram for sulfur (S); for illustrative purposes, we will simplify the S chemistry somewhat. (A complete pe–pH

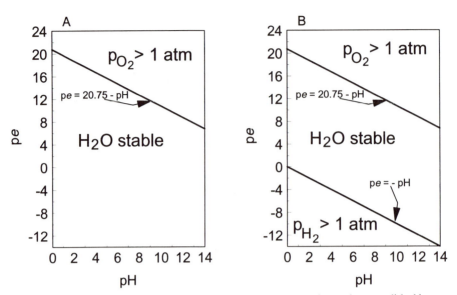

Figure 2.6. The construction of the H_2O pe–pH stability diagram. The diagram is accomplished in two simple steps. In the first step, we draw the line describing the equilibrium of H_2O with 1 atm of O_2 from Equation (2.7.3) (**A**). Above this line, conditions are too oxidizing and H_2O will spontaneously decompose into gaseous O_2; below the line, H_2O is stable with respect to O_2. In the second step, the line describing the equilibrium of H_2O with 1 atm of H_2 from Equation (2.7.5) is plotted (**B**). Below this second line, conditions are too reducing and H_2O will spontaneously decompose into gaseous H_2; above the line, H_2O is stable with respect to H_2. Thus the region of thermodynamic stability of H_2O exists between the two lines derived from Equations (2.7.3) and (2.7.5). One important conclusion from this diagram is that because aqueous-phase geochemistry requires the existence of water solutions, all processes in aqueous-phase geochemistry are necessarily constrained to occur within this region of pH and pe values.

stability diagram for S will be presented in Chapter 5, where we analyze the biogeochemical cycle of this element.) In our simplified S system, we only consider two oxidation states: –2 (i.e., sulfide) and +6 (i.e., sulfate). The relevant equilibria and their equilibrium constants for this system are four acid–base equilibria:

$$H_2SO_4 \leftrightarrows HSO_4^- + H^+ \qquad (\log K_{2.20} = 1.98) \qquad \text{(R2.20)}$$

$$HSO_4^- \leftrightarrows SO_4^{2-} + H^+ \qquad (\log K_{2.21} = -1.98) \qquad \text{(R2.21)}$$

$$H_2S \leftrightarrows HS^- + H^+ \qquad (\log K_{2.22} = -7) \qquad \text{(R2.22)}$$

$$HS^- \leftrightarrows S^{2-} + H^+ \qquad (\log K_{2.23} = -12.9) \qquad \text{(R2.23)}$$

and the redox half reaction that couples sulfate to sulfide:

$$SO_4^{2-} + 8H^+ + 8e^- \leftrightarrows S^{2-} + 4H_2O \qquad (\log K_{2.24} = 20.74) \qquad \text{(R2.24)}$$

Because of the existence of the different conjugate acids and bases in this system, the construction of the pe–pH stability diagram for S (even for this simplified version of S) is considerably more complicated than the H_2O stability diagram constructed previously. As a first step, we note that because the system we are considering is an aqueous one, we need only consider the values of pH and pe that are contained within the stability field of H_2O. Thus the first set of lines to draw in our diagram are simply those from the H_2O stability diagram derived previously (see Figure 2.8).

The next step is to consider the relevant acid–base equilibria; specifically we need to locate the locus of points where each acid and its conjugate base are equal in concentration. Recall from our discussion in Section 2.3 that this always occurs when pH = pK_a, where K_a is the appropriate acid–base equilibrium constant. [The reader may wish to review Equation (2.3.9) and the accompanying discussion at this point.] Thus

$$[H_2SO_4] = [HSO_4^-] \qquad \text{when pH} = -1.98 \qquad (2.7.6a)$$

$$[HSO_4^-] = [SO_4^{2-}] \qquad \text{when pH} = 1.98 \qquad (2.7.6b)$$

$$[H_2S] = [HS^-] \qquad \text{when pH} = 7 \qquad (2.7.6c)$$

and

$$[HS^-] = [S^{2-}] \qquad \text{when pH} = 12.9 \qquad (2.7.6d)$$

As can be seen in Figure 2.8B, the boundaries described by Equations (2.7.6) are lines of constant pH that appear on our pe–pH stability diagram as simple vertical lines running parallel to the y-axis. Note also in Figure 2.8B that while Equations (2.7.6) describe four vertical boundaries, only three vertical lines appear in Figure 2.6.4B. Because a pH < 0 is not encountered in natural waters, the minimum value

Figure 2.7. Thermochemical properties of natural waters. The range of pHs and pe values typical of different types of natural water systems suggests that a rich and diverse array of acidities and oxidation states are available for geochemical processes to occur.

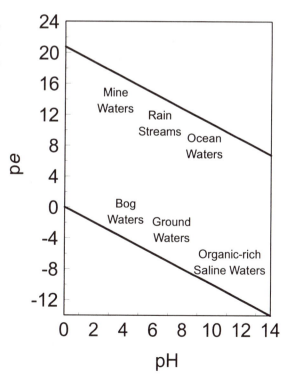

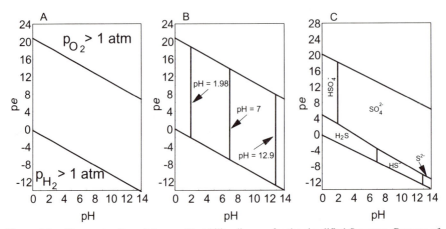

Figure 2.8. The construction of the pe–pH stability diagram for the simplified S system. Because of the existence of acid–base equilibria in this system, the construction of its stability diagram is more complicated than that for the H$_2$O stability diagram. As a first step in this process, we draw the stability lines for H$_2$O **(A).** Because all the species in our system are aqueous-phase species dissolved in water, they are constrained to reside within the stability field of H$_2$O drawn here. In the next step, we draw the constant lines of pH from Equation (2.6.6) that describe the boundaries between the relevant acids and their conjugate bases **(B).** In the final step, we draw the four lines from Equation (2.6.7) over the appropriate pH ranges that describe the boundaries between the S species in the $+6$ and -2 oxidation states (Panel C). As we do this step we also do the following: Trim the constant pH lines drawn in Panel B so that the pH line separating HSO$_4{}^-$ and SO$_4{}^{2-}$ is in the upper portion of the diagram where the $+6$ oxidation species reside and the pH lines separating the -2 oxidation state species are in the lower portion of the diagram where these species reside, and write in the names of the predominant species in each of the five pe–pH regimes defined by the lines we have drawn. In doing this, we need to keep in mind that oxidized species predominate in the upper portions of the diagram and acidic species predominate in the left-hand side of the diagram. It is interesting to note that although four separate lines were drawn in Panel C—one between $0 < \text{pH} < 1.98$, another between $1.98 < \text{pH} < 7$, another between $7 < \text{pH} < 12.9$, and another for $\text{pH} > 12.9$—they match up at their ends so that they form one continuous line with slight variations in slope as we move from one pH regime to another. This provides a useful check on the accuracy of the results. If you are constructing a pe–pH stability diagram and the combination of lines you have drawn that separate the species in one oxidation state from another is discontinuous, it means you have made an algebraic mistake and need to rederive your equations.

of pH along the x-axis of our diagram has been assigned a value of 0 and the boundary between H$_2$SO$_4$ and HSO$_4{}^-$ at pH $= -1.98$ does not appear.

The final step is to consider the redox equilibria. Fortunately, this task is simplified somewhat by our derivation of Equations (2.7.6). In addition to establishing the boundaries between the acids and their conjugate bases, these equations also let us know which redox equilibria we need to consider at various values of pH. For instance, for $0 < \text{pH} < 1.98$ the principal S species in the $+6$ oxidation state will be HSO$_4{}^-$ and the principal species in the -2 oxidation state will be H$_2$S. Thus in this pH region, we need to locate the line that describes the conditions where [HSO$_4{}^-$] = [H$_2$S]. From the equilibrium relationships for Reactions (R2.21), (R2.22), (R.23), and (R2.24), we find that

$$[\text{HSO}_4{}^-] = [\text{H}_2\text{S}] \qquad \text{when p}e = 4.83 - \frac{9}{8}\,\text{pH} \qquad (2.7.7\text{a})$$

Similarly for $1.98 < \text{pH} < 7$, we need to determine when SO_4^{2-} and H_2S are equal in concentration. Consideration of the appropriate equilibrium relationships reveals that

$$[SO_4^{2-}] = [H_2S] \quad \text{when } pe = 5.08 - \frac{10}{8} \text{pH} \qquad (2.7.7b)$$

For $7 < \text{pH} < 12.9$, the relevant equilibrium is between SO_4^{2-} and HS^- and we find that

$$[SO_4^{2-}] = [HS^-] \quad \text{when } pe = 4.21 - \frac{9}{8} \text{pH} \qquad (2.7.7c)$$

Finally for $\text{pH} > 12.9$, we must consider the equilibrium between SO_4^{2-} and S^{2-}:

$$[SO_4^{2-}] = [S^{2-}] \quad \text{when } pe = 2.6 - \text{pH} \qquad (2.7.7d)$$

Plotting these four equations on the pe–pH diagram establishes the boundaries between $+6$ and -2 S and yields a completed stability diagram (see Figure 2.8). By following these three steps you should, in principle, be able to construct pe–pH stability diagrams for any element. However, there is one additional complication that can arise from time to time that we should address. This complication—metastable boundaries—is discussed below.

2.7.3. Metastable Boundaries in pe–pH Stability Diagrams

Let us consider a somewhat more complicated S chemistry than that treated above, a system that includes three oxidation states for S: the $+4$ oxidation state (i.e., sulfite), as well as the -2 and $+6$ oxidation states. Following the procedure described above and the thermochemical data in the Appendix, we can now derive equations describing the boundaries between S in the $+6$ and $+4$ oxidation states and the $+4$ and -2 oxidation states. Plotting these equations in pe–pH phase space along with Equation (2.7.7) derived in Section 2.7.2 for the boundary between the $+6$ and -2 oxidation states, we obtain the three lines illustrated in Figure 2.9.

The two new lines in Figure 2.9 (i.e., the lines describing the boundaries between the $+6$ and $+4$ oxidation states and the $+4$ and -2 oxidation states) present somewhat of a conundrum. Consider, for example, the boundary between the $+4$ and -2 oxidation states (labeled "A" in Figure 2.9). This line describes the region of pe and pH where S($+4$) and S(-2) are in equal abundance. However, note that this line lies above the region where we find the boundary between the $+6$ and -2 oxidation states (labeled "B" in the figure). In other words, above line "B" S($+6$) is the most stable form of S. Because the upward direction in a pe–pH diagram represents movement toward more oxidizing conditions and S($+6$) is more oxidized than S($+4$), it follows that S($+6$) must be the most stable form of S in the region of pe–pH phase space where the boundary between S($+4$) and S(-2) is found. Hence, line "A" represents a boundary between two oxidation states of S in a region where neither is the most stable form of S.

A similar situation applies to the boundary between S($+6$) and S($+4$) (labeled

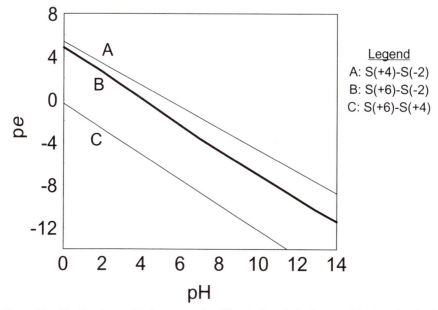

Figure 2.9. The lines in pe–pH phase space describing the boundaries between S in the $+4$ and -2 oxidation states (A), the $+6$ and -2 oxidation states (B), and the $+6$ and $+4$ oxidation states (C). Lines A and C represent *metastable boundaries* and thus can be eliminated from the S pe–pH stability diagram. Line A lies above line B and hence is in a region where S($+6$) is the most stable form of S and line C lies below line B and hence is in a region where S(-2) is the most stable form of S.

"C" in Figure 2.9). In this case we find that this line lies below line "B", in a more reducing region of pe–pH phase space. Hence in the region of the S($+6$)$-$S($+4$) boundary, S(-2) is the most stable form of S.

Lines in a pe–pH diagram, such as "A" and "C" in Figure 2.9, that represent boundaries between two oxidation states of an element in a region of pe and pH where neither oxidation state is the most stable form of the element are referred to as *metastable boundaries*. These boundaries do not represent true transitions between stable forms of the element and thus can be eliminated from the pe–pH stability diagram. In the specific case of the S system considered here (involving S in the $+6$, $+4$, and -2 oxidation states), we would eliminate lines "A" and "C" from the diagram. And, after adding in the lines describing the stability field of H_2O and the lines describing the relevant acid–base equilibria, we would obtain the same pe–pH stability diagram presented in Figure 2.8C.

2.8. CONCLUSION

Chemical thermodynamics allows us to quantitatively describe the equilibrium speciation of any chemical system provided we have the necessary thermochemical data. Of particular interest to us has been four general types of equilibria arising from acid–base reactions, gas-phase/aqueous-phase transitions,

TABLE 2.4
Thermochemical Equilibrium Relationships

Process	Reaction	Equilibrium Equation
Generic	$aA + bB \rightleftarrows cC + dD$	$K = [\{C\}^c\{D\}^d]/[\{A\}^a\{B\}^b]$ $= \exp\{-\Delta G^0/RT\}$
Acid–base	$HA \rightleftarrows H^+ + A^-$	$K_a = [H^+][A^-]/[HA]$:
Gas-phase dissolution	$X_g \rightleftarrows X_{aq}$	$K_{H,X} = [X_{aq}]/px$
Solid dissolution	$AB_S \rightleftarrows A^+ + B^-$	$K_{S,AB} = [A^+][B^-]$
Redox half-reaction	$OX + ne^- \rightleftarrows RED$	$K_{e,OX} = [RED]/[OX][e^-]^n$
Redox reaction	$OX_1 + RED_2 \rightleftarrows RED_1 + OX_2$	$K = K_{eOX1}/K_{e,OX2}$ $= [OX_2][RED_1]/[RED_2][OX_1]$

aqueous-phase/solid-phase transitions, and redox reactions. A listing of each of these four processes and the nomenclature we are adopting here to describe the equilibrium equations is presented in Table 2.4. You can use this table as a quick reference to supplement the material in the later chapters whenever necessary.

Before leaving our discussion of chemical thermodynamics, it is important to note a significant limitation of the relationships we have derived in this chapter. These relationships are equilibrium relationships and nothing more. They allow us to make only two inferences: (1) the characteristics of the equilibrium state of any given system and (2) the fact that a system not in equilibrium will spontaneously tend toward that equilibrium. One thing that chemical thermodynamics specifically do not prescribe is how rapidly a system will move to its equilibrium state. (The rate of equilibration is governed by another branch of chemistry, namely, chemical kinetics.) Moreover, the rate at which a given chemical system moves toward its equilibrium state can, in principle, be quite slow. For this reason, it is not uncommon to find natural systems that are far from their equilibrium states. On the other hand, because these systems must be moving, albeit slowly, toward equilibrium, we can at least use chemical thermodynamics in these cases to determine the general direction these systems will tend to follow.

SUGGESTED READING

Fischer, R. B., and D. G. Peters, *Chemical Equilibrium*, W. B. Saunders, London, 1970.

Mahan, B. H., *University Chemistry*, Addison-Wesley, Reading, MA, 1980.

Hobbs, P. V., *Basic Physical Chemistry for the Atmospheric Sciences*, Cambridge University Press, New York, 1995.

Stumm, W., and J. J. Morgan, *Aquatic Chemistry: An Introduction Emphasizing Chemical Equilibria in Natural Waters*, John Wiley & Sons, New York, 1981.

PROBLEMS

1. Oxygen in peroxides (e.g., H_2O_2) has as an anomalous oxidation state of -1. In all other molecules, oxygen has an oxidation state of -2, or, in the case of oxygen allotropes, such as O_2 and O_3, an oxidation state of 0. Look up in a chemistry textbook or reference book the chemical structure of hydrogen peroxide. Use this structure to explain why the oxygen atoms have an oxidation state of -1 on the basis of this structure.

2. Derive the pH and total concentration of ammonia in a liquid water solution in equilibrium with ammonia gas at a partial pressure of 1×10^{-6} atm, 1×10^{-4}, and 0.01 atm.

3. Use the data in the Appendix and the methodology discussed in this chapter to reproduce lines "A" and "C" in Figure 2.9.

4. Determine the locations of the stability boundaries of H_2O_2 in pe–pH phase space relative to O_2 and to H_2O.

The Earth System

3

"The Earth is like a four-ring circus with acts peculiar to each ring going on simultaneously. . . . Unlike the rings of a circus [however], the four Earth "spheres" are not isolated from each other. One influences the other and at times the boundaries are not easily discerned."

K. K. Turekian, *Oceans*, Prentice-Hall, Englewood Cliffs, New Jersey, 1976.

3.1. INTRODUCTION

In Chapter 2 we reviewed chemical thermodynamics and the types of chemical transformations that an element can undergo in a biogeochemical cycle. However, chemical transformations are not the only kinds of changes at work in biogeochemical cycles. There are also geophysical, atmospheric, and hydrospheric processes at work on the earth that cause elements to move from one physical location to another. In most cases these two types of transformations—chemical and physical—are highly coupled. For instance, a chemical transformation that increases an element's mobility, by converting it from a solid to a liquid or gas, may also facilitate its geological movement from one location on the earth to another. Conversely, the transport of an element to a new environment having significantly different thermochemical properties will likely also trigger a change in the thermochemical properties of the element via, for instance, a redox reaction. To help us understand this close coupling between chemical transformation and physical transportation in biogeochemical cycles, we need to first develop a working knowledge of the various kinds of physical and chemical environments found on the earth. Toward that end, we review in this chapter the essential facets of the *earth system*.

To even the most casual observer, the earth is not a homogeneous body, but a composite of different domains having distinctly different physical and chemical properties. Probably at the most obvious level, the earth can be divided into three domains: the atmosphere, the oceans, and the solid earth. Although often less apparent, other subdivisions of the earth can be inferred. For instance, by studying the

41

properties of *seismic waves*, geophysicists have determined that the earth's interior actually consists of three layers: the core at the earth's center, the mantle which surrounds the core, and the crust at the earth's surface. Because the properties of the crust vary depending upon whether it lies underneath a continent or an ocean, it is often further subdivided into continental crust and oceanic crust. Together these six domains—the atmosphere, ocean, continental and oceanic crusts, mantle, and core—comprise the major physical components of the earth system. A brief summary of the properties of each of these domains is presented in Table 3.1 (see also Text Box 3.1).

Although the six domains described above provide a useful framework for classifying different physical components of the earth, they are by no means the only classification scheme for the planet. In the specific case of biogeochemical cycles, it is useful to adopt a scheme that places greater emphasis on the role of living systems and less emphasis on the inner layers of the solid earth. For this reason, biogeochemists often adopt a classification scheme that divides the earth into four domains or *spheres*. Two of these are the same as in our earlier scheme: they are the *hydrosphere*, which consists of the ocean and all fresh waters, and the *atmosphere*. The third domain, called the *lithosphere*, consists of the first 50–100 km of the solid earth and includes the crust and the uppermost portion of the mantle; because biogeochemical cycles rarely involve exchange of material with the deeper mantle and core, these portions of the earth system are usually not included in studies of the global biogeochemical cycles. The fourth domain is comprised of the collection of all living organic material on the earth and is referred to as the *biosphere*. As can be seen from an inspection of Figure 3.1, these four regimes represent unique chemical as well as physical regimes with distinctly different elemental makeups.

3.2. THE HYDROSPHERE

The hydrosphere consists of all the water on the earth. This includes the water contained in lakes and rivers, the frozen water in glaciers and polar ice, and all ground water, as well as the oceans. However, the oceans comprise more than 97% of the earth's water, and thus we can reasonably limit our discussion of the hydrosphere to the oceans.

TABLE 3.1
The Major Subdivisions of the Earth

Subdivision	Thickness (km)	Mass (g)	Mean Density (g cm^{-3})
Core	3480	2×10^{27}	10.6
Mantle	2870	4×10^{27}	4.6
Oceanic crust	7	7×10^{24}	2.9
Continental crust	40	2×10^{25}	2.75
Ocean	4	1.4×10^{24}	1
Atmosphere	Undefined	5×10^{21}	Variable
Earth	6371	6×10^{27}	5.5

Source: J. C. G. Walker, *Evolution of the Atmosphere*, Macmillan, New York, 1977.

TEXT BOX 3.1	THE ATMOSPHERE, OCEAN, AND SOLID EARTH: GAS, LIQUID, AND SOLID?

The division of the earth into an atmosphere, ocean, and so-called solid earth would appear to imply that the earth has neatly divided into three separate phases: a gaseous atmosphere, a liquid ocean, and a solid earth. In reality, the phases do not separate quite so neatly on our planet. As we all know from walking around on a rainy day, it is not uncommon to find liquid or even frozen (i.e., solid) water in the atmosphere. Moreover, a portion of the ocean in the polar regions is permanently frozen; this is often referred to as the *cryosphere*. And the outer portion of the core is actually molten; in other words, it is a liquid.

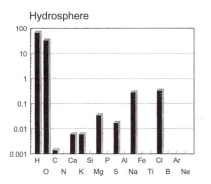

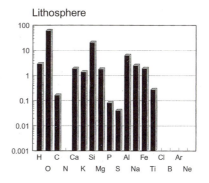

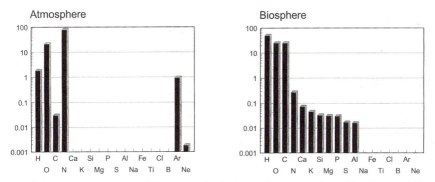

Figure 3.1. Percent abundance of the elements in the hydrosphere, atmosphere, lithosphere, and biosphere. (After E. S. Deevey, Mineral cycles, in *The Biosphere, A Scientific American Book*, W. H. Freeman, San Francisco, 1970.)

3.2.1. The Ocean Bottom

The world's oceans comprise about 0.02% of the earth's mass but cover about 70% of the earth's surface, with roughly two times as much ocean water residing in the Southern Hemisphere than in the Northern Hemisphere (see Figure 3.2). The oceans are bounded by the atmosphere from above and the deep-sea floor and continents from below and from the sides. As illustrated in Figure 3.3, there is usually a gradual transition region between the deep-sea floor or basin and the continents spanning about 1500 km; this transition region, called the *continental margin*, can be further subdivided into the *continental shelf, slope,* and *rise*. The deep-sea floor itself has an average depth of about 4 km; however, the sea floor is characterized by large variations in topography, with (a) deep depressions or *trenches* (having maximum depths of as much as 11 km) and (b) elevations called *ridges, seamounts,* and *guyouts* that can rise as much as 2.5 km above the basin floor.

3.2.2. Physical Properties of the Ocean

Variations in ocean temperatures as a function of depth, such as that illustrated in Figure 3.4, indicate the existence of three distinct ocean layers. At the top, we

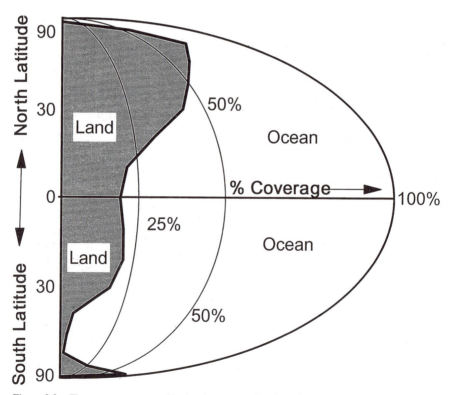

Figure 3.2. The percent coverage of land and ocean as a function of latitude. For the entire globe, about 70% of the earth's surface is covered by ocean, with about twice as much in the Southern Hemisphere as in the Northern Hemisphere. (After Garrels, R.M., F.T. Mackenzie, and C. Hunt, *Chemical Cycles and the Global Environment*, William Kaufman, Los Altos, California, 1975.)

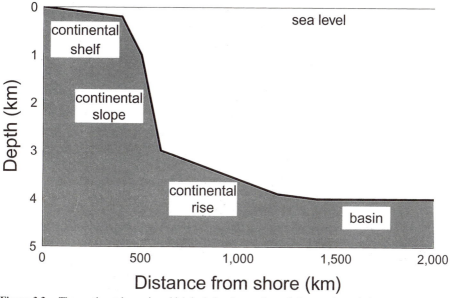

Figure 3.3.   The continental margin, which includes the continental rise, continental slope, and continental rise, represents the transition region between the ocean basin and the continent.

generally find a well-mixed surface layer of about 50–200 m in depth with little or no vertical variation in temperature. Below the surface layer, there is usually a region of about 500–1000 m in depth where temperature falls off rapidly; this region is referred to as the *thermocline*. Finally, below the thermocline, we encounter the *deep ocean* where temperatures slowly decrease with depth to a minimum temperature near the ocean bottom of about 2°C. For the most part, in our studies of global biogeochemical cycles, we will find it convenient to simplify these divisions somewhat by combining the deep ocean and the thermocline into a single reservoir which we will refer to, for simplicity, as the deep ocean.

The ocean circulation can be thought of as having two components. The circulation of the surface layer is largely a *wind-driven circulation* with motion principally in horizontal directions. Because of the effects of the earth's rotation (i.e., the *Coriolis effect*), this circulation is dominated by large-scale gyres moving clockwise in the Northern Hemisphere and counterclockwise in the Southern Hemisphere. The circulation of the deep ocean involves both vertical and horizontal movement and is driven by variations in the densities of the various masses of water moving about in the deep ocean. Because the density of ocean water is determined by its temperature and salinity, the circulation of the deep ocean is referred to as a *thermohaline circulation*. Two of the most important water masses found in the deep ocean are the North Atlantic Deep Water and the Antarctic Bottom Water. The North Atlantic Deep Water originates from winter-cooled salty water in the surface layer of the North Atlantic Ocean near Greenland; this water sinks to depths of about 1.5–4 km and flows southward into the South Atlantic. The Antarctic Bottom Water originates in the Weddel Sea in the Antarctic, sinks to depths of 3 km or more, and flows northward.

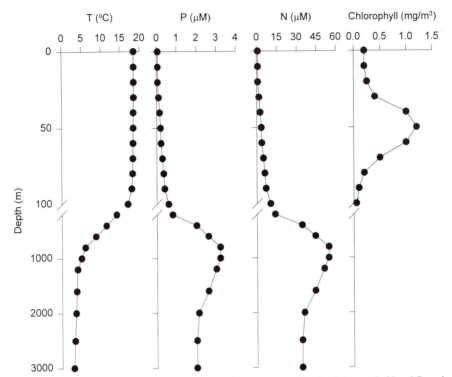

Figure 3.4. Typical vertical distributions of oceanic temperature, dissolved, inorganic N and P, and chlorophyll (an indication of the presence of photosynthesizing phytoplankton). Note the separation of the ocean into a well-mixed surface layer (with relatively constant temperatures, high levels of photosynthetic activity, and depleted concentrations of nutrient elements), a deep ocean (with cooler temperatures, no photosynthetic activity, and higher concentrations of nutrient elements), and a transition zone or thermocline between the surface layer and deep ocean where temperatures fall rapidly with depth. In this figure, the thermocline is found between about 100 and 800 m. The euphotic zone, where light from the sun is strong enough to support photosynthetic activity, generally extends to the depth of maximum chlorophyll concentration—about 50 m in this figure. In general, the depth of both the thermocline and euphotic zone vary depending upon season and latitude, atmospheric conditions, and ocean turbulence and turbidity.

3.2.3. Ocean Chemistry

From a chemical point of view, ocean water is a briny aqueous solution containing, on average, about 3.5% salt by weight. This ocean salinity arises primarily from salts of the chloride ion (Cl^-) and, to a lesser extent, the sulfate ion (SO_4^{2-}), with the major cations being, in order of abundance, sodium (Na^+), magnesium (Mg^{2+}), calcium (Ca^{2+}), and potassium (K^+). Thus as indicated in Figure 3.1, the major elements of the ocean are hydrogen (H) and oxygen (O), from water, as well as Cl, Na, Mg, S, Ca, and K.

In addition to the major elements, the ocean contains an array of more variable trace elements. Among the trace elements of the ocean, the one's of particular interest for us are the so-called *nutrient elements*. These elements include carbon (C), nitrogen (N), and phosphorus (P) (three of the five elements we will focus upon here), as well as silica (Si), iron (Fe), and many others. The nutrient elements are required by *phytoplankton*, the free-floating, microscopic green plants of the ocean, to carry out photosynthesis. Since phytoplankton occupy the bottom of the ocean's food chain, the organic material produced by these organisms ultimately provides the food supply for all of the ocean's other creatures. For this reason, the distribution of the nutrient elements in the ocean is critically important to the biogeochemistry of the ocean.

Because the nutrient elements are required for photosynthesis, the oceanic distributions of these elements are strongly impacted by biological processes. Consider, for instance, the vertical distributions of nitrogen, phosphorus, and chlorophyll (a measure of *phytoplankton* abundance) illustrated in Figure 3.4. The strong anticorrelation between chlorophyll and the nutrient elements, nitrogen and phosphorus, in this figure is quite typical and is directly related to the precarious balance between life and death for the ocean's phytoplankton. Because photosynthesis requires sunlight and sunlight is rapidly attenuated as it passes through ocean water, the photosynthetic activity of phytoplankton is limited to the top 50–100 meters of the ocean— the so-called *euphotic zone*. Since phytoplankton must photosynthesize to survive, they can only live within this euphotic zone. Unfortunately for the phytoplankton, they are denser than water and tend to sink, and, despite having a variety of adaptive strategies, they eventually sink below the euphotic zone and die. Thus, the ocean surface layer is relatively rich in phytoplankton (and the chlorophyll from these organisms), and the deep ocean is essentially devoid of phytoplankton. Because phytoplankton consume nutrient elements in the process of photosynthesis, the concentrations of these elements are relatively low in the surface layer. A maximum in nutrient element concentrations is usually found in the upper portion of the deep ocean (about 800–1000 m in Figure 3.4), where dead and sinking organic matter decays or is *mineralized*. The nutrient elements from which the organic matter was originally made is returned back to the deep ocean waters by the decay of the organic matter and, in the process, gives rise to an ocean cycle for the nutrient elements that ultimately makes photosynthesis by phytoplankton, and hence all of ocean life, possible (see Figure 3.5 and Text Box 3.2).

Although biological processes can affect the distribution of nutrient elements, these very same biological processes can, in turn, be affected by the distribution of nutrient elements. Because nutrient elements are in short supply in the ocean surface layer, those regions where the sources of nutrient elements are relatively large generally tend to be regions of relatively intense photosynthetic activity as well. Within the ocean surface waters, there are two major sources of nutrients: *ocean upwelling* and *river runoff*. The term *upwelling* is used to describe the process by which ocean deep water is brought to the surface or upwells. Upwelling is driven by atmospheric winds and ocean currents, in conjunction with the centrifugal effects of the earth's rotation, and occurs most intensely along the eastern shores of the world's oceans—for example, the Pacific coast of North and South America and the Atlantic coast of Africa. Because ocean deep water is relatively rich in nutrients, the upwelling of this water represents a source of nutrients for the surface layer. Similarly, because fresh water rivers and streams tend also to be rich in nutrients, the flow or runoff of these waters into the ocean represents a nutrient source to the

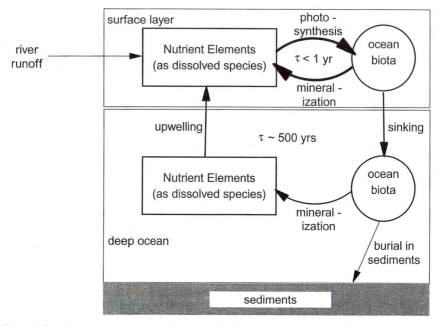

Figure 3.5. The ocean-nutrients cycle. Squares and circles represent reservoirs of the nutrient elements, arrows represent fluxes of the nutrients from one reservoir to another, and "τ" indicates the approximate time scale for the nutrient elements to traverse a complete cycle. Note that the ocean-nutrients cycle actually consists of two subcycles—a relatively short surface layer cycle and a longer cycle involving exchange between the deep ocean and the surface layer (see Text Box 3.2).

surface layer as well. Both of these sources are most intense near the continents, and it is for this reason that the most (photosynthetically) productive regions of the ocean are generally found along the continental shelves (see Figure 3.6).

3.2.4. The Age of the Oceans—A Paradox?

We conclude our brief discussion of the hydrosphere by presenting a paradox. The oceans cover approximately 70% of the earth's surface and have an average depth of about 4000 m. From these numbers, and the fact that the surface area of the earth is equal to 5×10^{14} m^2, we can estimate that the oceans occupy a volume of about 1.4×10^{18} m^3. This volume of water is maintained over time by an approximate balance between the sources of ocean water from rivers and rain and the loss of water due to evaporation (see Table 3.2). However, in addition to water, the rivers bring other material to the ocean. This material comes from the continents and is generated from the erosion or *weathering* of the soils and rocks that make up the continents. As noted above, some of this other material in river water is in the form of dissolved nutrients, whereas other material is in the form of undissolved particles, such as clays and silt.

It is estimated that erosion lowers the continents by about 6 cm every 1000 years. This is equivalent to the loss of about 10^{10} m^3 of continental material each year. Where does this volume of material go? To a first approximation, it is carried away

TEXT BOX 3.2	THE CYCLE OF OCEAN NUTRIENTS

The removal of the nutrient elements from the surface layer of the ocean by phytoplankton during photosynthesis occurs at a rapid rate. In fact, it is estimated that if the organic matter synthesized by phytoplankton were to be stored permanently in that form, photosynthesis would deplete the surface layer of the ocean of all its nutrient elements in less than one year, rendering further photosynthesis impossible and causing the oceans to die. This of course does not happen because there is another process: respiration and decay. Respiration and decay mineralizes the nutrient elements from organic matter and returns them to the water column in dissolved form so that they may participate again in the photosynthetic process (see Figure 3.5). About 90% of the organic matter produced in the surface layer by photosynthesis is mineralized within the surface layer, where the freshly released nutrient elements can be immediately reutilized in photosynthesis. However, a small fraction of the organic matter escapes mineralization in the surface layer and sinks to the deep ocean. Mineralization of this organic matter produces nutrient elements, but in a region of the ocean where they cannot be utilized by phytoplankton in photosynthesis. The characteristic time for ocean upwelling to recycle the nutrients from the deep ocean to surface layer is quite long—of the order of 1000 years. It is also worth noting that the ocean nutrient cycle is an "open cycle" (as defined in Chapter 1), in the sense that there are nonzero flows of nutrients into and out of the surface layer and deep ocean reservoirs that define the cycle. The loss of nutrients arises from the fact that about 0.1% of the nutrient elements taken up by phytoplankton in photosynthesis escapes mineralization in both the surface layer and the deep ocean and is buried in the sediments at the ocean bottom. This loss of nutrients is approximately balanced by the influx of nutrients to the ocean from river runoff and, to a lesser extent, atmospheric precipitation.

by the rivers to the oceans. Being largely nonvolatile, this material must eventually find its way to the bottom of the ocean where it forms ocean sediments. However, we just estimated that the volume of the ocean is only 1.4×10^{18} m^3. The dumping of 10^{10} m^3 of continental material into the oceans each year would therefore fill the entire ocean with mud and rocks in about 140 million years. Clearly the oceans are not full of mud and rocks, so this would imply that the oceans are younger than 140 million years. Interestingly, this time frame of 100–200 million years is consistent with the ages measured for the sediments that form at the bottom of the ocean. The oldest of these sediments is estimated at about 150 million years. The only problem is that the earth is about 4.9 billion years old and there is strong geologic evidence that the oceans appeared no later than 3.7 billion years ago. Moreover, we have fossil evidence for the existence of shell-forming invertebrates in the ocean some 500–600 million years ago. How can we resolve these contradictory pieces of evidence, some pointing to a young ocean (only about 140 million years old) and

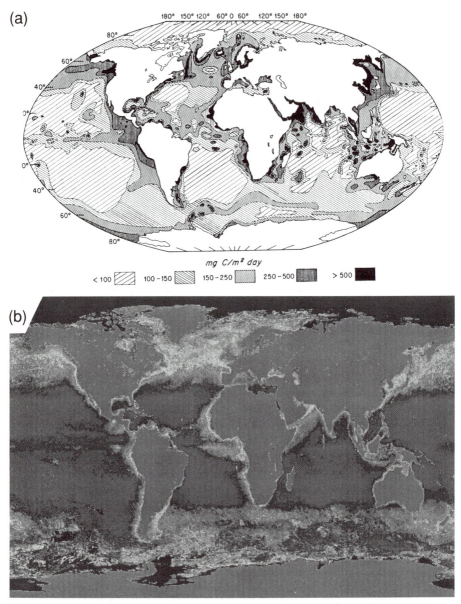

Figure 3.6. Two pictures of the distribution of ocean productivity. Panel A: the distribution of primary production in the world's ocean developed by Russian scientist Koblentz-Mishke and colleagues in the 1960's from *in-situ* measurements. The picture appearing here was modified by Degens and Mopper (in *Chemical Oceanography*, J. P. Riley and R. Chester (eds.), 2nd ed., **6**, Academic Press, New York, 1976, pp. 60–113.) and provided courtesy of K. Mopper. Panel B: The distribution of ocean productivity inferred from the Ocean Color Experiment, a space-based measurement using remote sensing. In the picture, the lighter areas of the ocean indicate regions of relatively high productivity and the darker areas, low productivity. Provided courtesy of the National Aeronautics and Space Administration.

TABLE 3.2
The Ocean Water Cycle

Reservoir of ocean H_2O	1.4×10^{18} m^3
Ocean sources	
River runoff	4.0×10^{13} m^3 year^{-1}
Precipitation	3.9×10^{14} m^3 year^{-1}
Ocean loss	
Evaporation	4.3×10^{14} m^3 year^{-1}

Source: W. Schlesinger, *Biogeochemistry—An Analysis of Global Change*, Academic Press, New York, 1991.

others pointing to an old ocean (billions of years in age)? The answer can be found in the lithosphere and plate tectonics.

3.3. THE LITHOSPHERE

As discussed earlier, the vast majority of the earth's mass is found in the solid earth, and the solid earth consists of three separate layers: the core, the mantle, and the continental and oceanic crusts (see Table 3.1). The earth's core is composed primarily of iron (Fe) with about 10–20% nickel (Ni) in the solid, inner portion of the core and about 10% sulfur (S) and small amounts of silicon (Si) and potassium (K) in the liquid, outer portion of the core. The mantle is composed of a mixture of iron and magnesium silicates called *peridotite*. By contrast to the core and mantle, the crust is a much more heterogeneous mixture of rocks with differing chemical compositions. In general, however, we can say that the crust tends to be richer in Si and aluminum (Al) and poorer in Fe and Mg than the mantle, with continental crust being more so than oceanic crust. (Geologists use the term *sialic* to describe material like the crust that is rich in Si and Al and use the term *mafic* to describe material like the mantle that is rich in iron and magnesium.) As noted earlier, the lithosphere includes the crust and the upper portion of the mantle and encompasses the outer 100 km of the solid earth. The reason for combining the crust and upper portion of the mantle into a single sphere will become apparent in Section 3.3.2 when we discuss the theory of plate tectonics. But first, we briefly review the kinds of rocks and minerals that make up the earth's crust.

3.3.1. Rocks and Minerals of the Crust

The crustal portion of the lithosphere is of particular interest because it is the portion that interacts directly with the other spheres of the biogeochemical system. As noted above, the crust is essentially a heterogeneous mixture of different kinds of rocks. Rocks, in turn, are mixtures of minerals, and minerals are defined as crystalline solids having specific chemical compositions. Rocks are generally categorized by geologists according to the process by which they are formed: The categories are *igneous*, *sedimentary*, and *metamorphic* (see Table 3.3).

Igneous rocks are formed through the crystallization of molten magma. When the molten magma reaches the earth's surface (i.e., it becomes a lava flow), the re-

TABLE 3.3
The Rocks of the Crust

Types	Examples
Igneous Rocks	
Extrusive	Basalts (i.e., mafic)
Intrusive	Granites (i.e., sialic)
Sedimentary Rocks	
Detrital	Shale (from mud), sandstone (from sand)
Biological	Limestone (from shells, coral reefs), chert (from diatoms)
Evaporites	Gypsum ($CaSO_4 \cdot 2H_2O$), anhydrite ($CaSO_4$), halite ($NaCl$), calcite ($CaCO_3$)
Metamorphic Rocks	
	Slate (derived from shale)
	Marble (derived from limestone)

sulting rock is referred to as *extrusive* igneous rock. Conversely, when magma crystallizes beneath the earth's surface, it is referred to as *intrusive* igneous rock. Extrusive igneous rocks are usually derived from magma from the upper mantle and are therefore relatively rich in iron and magnesium (i.e., mafic). The most abundant mafic igneous rock is basalt, which is commonly found along the sea floor. Intrusive rocks, on the other hand, are usually derived from magma from the crust and thus are relatively rich in silica and aluminum (i.e., sialic). Granite is a common form of this type of igneous rock.

Sedimentary rocks are formed from the sedimentation and/or precipitation of material and its subsequent compaction. There are three types of sedimentary rocks. Detrital sedimentary rocks are derived from the sedimentation of solid material and include shale derived from mud and sandstone derived from sand. Biological sedimentary rocks are derived from dead organic material and include (a) limestone from shells and coral reefs and (b) chert from diatoms and radiolaria. Lastly, evaporites are sedimentary rocks derived from the physical precipitation of salts from ocean waters, often caused by the evaporation of shallow pools and ponds. Examples of evaporite deposits are gypsum ($CaSO_4 \cdot 2H_2O$), anhydrite ($CaSO_4$), halite ($NaCl$), and calcite ($CaCO_3$).

Metamorphic rocks have undergone some degree of chemical transformation as a result of being exposed to conditions intermediate between those of the surface and the high temperatures and pressures that cause melting into magma and the eventual formation of igneous rocks. These rocks include slate, derived from shale, and marble, derived from limestone.

3.3.2. Plate Tectonics

The theory of *plate tectonics* describes the mechanism by which the earth's continents and oceans continually move and rearrange themselves. The development and validation of this theory will almost certainly be looked upon as one of the most important, if not the most important, advance in the earth sciences during our century. However, the intellectual seeds from which the theory of plate tectonics sprouted in the twentieth century were actually sown long before our time. Speculation on the movement or drifting of the continents can be found in any number of pre-twentieth century treatises and manuscripts, including those of the seventeenth-century philosopher Sir Francis Bacon. Nevertheless, Alfred Wegener is generally given credit for formally advancing the hypothesis of *continental drift*. In his 1915 book entitled *The Origin of the Continents and the Oceans*, Wegener proposed that the continents were once joined together in a single supercontinent (that he called *Pangaea*) and that this mass began breaking into smaller continents and separating some 200 million year ago. Although Wegener based his hypothesis on a good deal of geological, paleontologcial, and climatological data, he lacked a plausible physical mechanism for explaining how continental drift could actually occur. Perhaps for this reason, and perhaps because his ideas ran so strongly counter to the accepted geological dogma of his time, Wegener's hypothesis was, by and large, dismissed by his scientific colleagues.

By the late 1960s and early 1970s, however, the perception of Wegener's ideas had changed radically. New data gathered from paleomagnetic measurements and oceanographic studies had established that the ocean floor was not a permanent or static feature but was in fact spreading outward from a worldwide system of *mid-ocean ridges*. The discovery of this phenomenon, known as *sea-floor spreading*, along with a growing body of data supporting the continental drift hypothesis of Wegener, finally forced the earth science community to embrace a new model of the solid earth—a model even more revolutionary than that of Wegener's drifting continents. This was a model of the continents as drifting plates.

In the theory of plate tectonics, the outer shell of the earth or lithosphere is a composite of some 20 separate rigid plates that move independently along the earth's surface under the influence of convective currents driven by geothermal energy from the earth's interior. The lithospheric plates overlie a more plastic or liquid portion of the mantle, called the *asthenosphere*, which provides a sufficiently "slippery" interface upon which the plates can move.

Geophysical measurements indicate that the plates move at a rate of a few to perhaps 10 cm each year. Although a shift of a few centimeters in land masses each year might seem like a small change, over geologic time scales this slow, but continuous movement of the plates brings about the tectonic upheavals that ultimately determine the topography of our planet (see Figure 3.7). Most dramatic of these upheavals are the earthquakes and volcanoes produced when two plates, moving in opposite directions, crash into each other along a *convergent plate boundary*. The resulting *subduction* of one plate under the other gives rise to magma, volcanic activity, and mountain building.

Alternatively, when two plates move away from each other along a *divergent plate boundary*, they produce a depression in the surface topography that becomes an ocean basin. At the nominal center of the ocean basin, along the boundary between the two separating plates, we find the *mid-ocean ridges*. These ridges are generally about 200 km in width and rise from 1 to 2.5 km above the ocean floor. At the center of the ridges is a rift zone where fresh magma from the mantle below

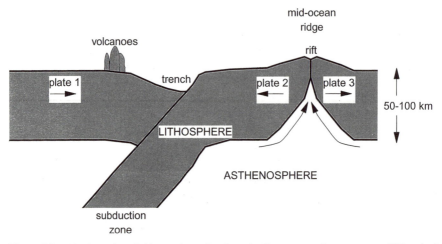

Figure 3.7. The formation of ridges and trenches from the divergence and convergence of lithospheric plates. The divergence of plates along the mid-ocean ridges gives rise to the formation of new oceanic crust and sea-floor spreading. The convergence of two plates causes the formation of a trench, subduction of one plate beneath the other, and tectonic activity. The topographical formations that result from plate convergence depends upon the type of convergent plate boundary. Along continental–oceanic plate boundaries, volcanic mountain ranges form on the continental plate; along oceanic–oceanic plate boundaries, volcanic island arcs form; and along continental-continental plate boundaries, mountain ranges form on the interior of the new combined continent. (After J. C. G. Walker, *Evolution of the Atmosphere*, Macmillan, New York, 1977.)

bubbles up to fill the void left behind by the receding plates. This magma cools and hardens into basaltic rock, attaching itself to one of the receding plates and effectively forming new sea floor (or oceanic crust) at the edge of the plates. The continual divergence of the plates causes this new sea floor to recede from the ridge, giving rise to the phenomenon known as *sea-floor spreading*. As the sea floor spreads, it accumulates sedimentary material from the detritus and precipitating minerals falling out of the ocean-water column. The further from the ridge the sea floor travels, the greater its age and, as a result, the deeper the layer of sedimentary material. Eventually the sea floor that had originally formed along the mid-ocean ridge, as well as the sedimentary material that had accumulated upon the sea floor, reaches the other edge of the plate where further movement causes it to collide with another oncoming plate. At this point it is subducted and brought back to the earth's surface in the mountain-building process. The result is a *rock cycle* that ensures a continuous supply of nutrient elements to the biosphere (see Text Box 3.3).

The beauty of the theory of plate tectonics has been its ability to solve a vast number of what had previously been unexplainable scientific puzzles about the earth. For instance, with plate tectonics we can resolve the paradox encountered in Section 3.2 on the age of the oceans and their fate. Recall that, given the rate of transfer of material from the continents to the oceans, we estimated that the oceans should fill up with mud and sediment in a relatively brief 140 million years. However, in light of plate tectonics, we now see that the oceans do not, in fact, fill up because the sea floor and the sedimentary material that accumulate upon it are constantly being recycled and returned to the continents. Similarly, the apparent inconsistency between the age of the sea floor (~100–200 million years) and the age of the oceans (billions of years) is not a problem. Because of sea-floor spreading and the recycling of the sea floor, we would expect the sea floor and the sedimentary material that accumulate upon it to be significantly younger than the oceans themselves.

TEXT BOX 3.3	THE ROCK CYCLE: THE FINAL LINK IN THE GLOBAL BIOGEOCHEMICAL SYSTEM

The ocean nutrient cycle discussed in Section 3.2 provides an efficient mechanism for recovering the nutrients lost from the surface layer in photosynthesis and making them available for repeated use by the biosphere. Nevertheless, the nutrient cycle is not perfect, and a small percentage (about 0.1%) of the nutrients are lost in each cycle and deposited on the ocean bottom as sediment. If left unchecked, this continuous loss of nutrients would slowly, but inevitably, deplete the ocean of all its nutrients, making further photosynthesis in the ocean impossible. Fortunately for life on earth, plate tectonics produces a rock cycle that keeps the system working. As illustrated in Figure 3.8, the motion of the plates brings the sedimentary rock formed at the ocean bottom along with the nutrient elements contained in them back to the earth's surface. Once at the surface, weathering and erosion mineralizes the nutrients, making them, once again, available to photosynthesizing biota. It is perhaps somewhat ironic that an abiotic process—plate tectonics—is ultimately responsible for keeping the biological part of the earth in operation. In fact, some planetary scientists think that life may have once existed on Mars, but could not be sustained because of the lack of tectonic activity. Without tectonic activity and a rock cycle, it is argued, the planet lacked a mechanism to recycle the nutrients needed to sustain life. (An excellent review of this subject can be found in a paper by J. F. Kasting, O. B. Toon, and J. B. Pollack in the February 1988 issue of *Scientific American*.)

3.4. THE ATMOSPHERE

Although the atmosphere contains less than one-millionth of the mass of the earth (see Table 3.1), it nevertheless plays an important role in biogeochemical cycles. Because the atmosphere is gaseous, and because the atmosphere is rapidly mixed by the winds and storms that make up our weather, it provides a ready medium for transporting elements from one location on the earth to another. In addition, as we describe below, it represents a unique chemical environment for interaction with the biosphere.

3.4.1. Atmospheric Composition

Approximately 99% of the atmosphere's mass is made up of two substances: molecular nitrogen (N_2) and molecular oxygen (O_2). In addition to these two substances, the atmosphere contains water (H_2O), argon (Ar), and carbon dioxide (CO_2),

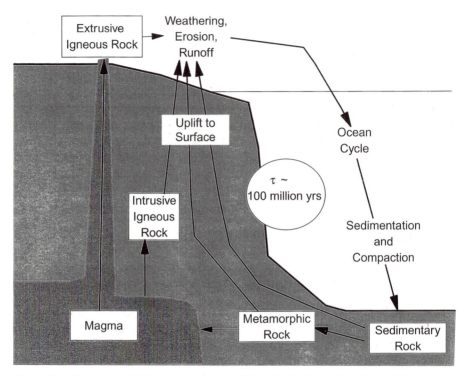

Figure 3.8. The rock cycle. Squares represent the different kinds of rocks found in the lithosphere, and the arrows represent the flow of material from one rock type to another. The rock cycle, through plate tectonics, returns nutrient elements that had been lost to ocean sediments back to the earth's surface where, via weathering, they become available to the biosphere. The time scale for an element to traverse the rock cycle is of the order of 100 million years (see Text Box 3.3).

as well as a myriad of trace species (both gaseous and particulate) present at concentration levels of a few parts per million or less (see Figure 3.9 and Table 3.4). Although their concentrations are quite small, the atmospheric trace species are of considerable concern because of their disproportionate impact on climate, ultraviolet solar radiation, and air quality.

The chemical composition of the atmosphere teaches us a good deal about its origin and its interaction with other elements of the earth system. For instance, consider the concentrations of the inert or rare gases, helium (He), neon (Ne), argon (Ar), krypton (Kr), xenon (Xe), and radon (Rn). With the exception of Ar (see Text Box 3.4), the concentrations of these gases are quite low. The concentrations are, in fact, much lower than one might expect given the relative abundances of these elements in the solar system and the fact that the earth is believed to have accreted from this same material. Because inert gases would not have been appreciably tied up in the solid earth when the planet accreted, these elements would have to have been in the earth's initial or primeval atmosphere when the planet first formed. The relative paucity of the inert gases in today's atmosphere therefore implies that they, along with the other contents of the primeval atmosphere, were stripped from the earth during the planet's early history, perhaps as a result of an intense solar wind or heavy bombardment by meteorites at the time.

If this is true, then where does our present atmosphere come from? The answer lies in plate tectonics and the rock cycle. The elements that make up our present-day atmosphere must have come from the earth's interior—that is, from the gases spewing into the atmosphere from the lithosphere in volcanoes and other venting processes. This *degassing* of volatile material from the lithosphere must have also supplied the water to the atmosphere that eventually condensed to form the world's oceans. Of course, the elements that make up the atmosphere and ocean are not statically tied to these spheres. Like the nutrient elements, they continuously move through the rock cycle, causing them to travel from the atmosphere and hydrosphere to the lithosphere and then back to the atmosphere and hydrosphere.

In addition to the role of the rock cycle, the chemistry of the atmosphere suggests that life has had a profound influence on our environment. Of all the planets of the solar system, the earth is the only one with such large amounts of free oxygen in the form of O_2. On Mars and Venus, for example, virtually all of the oxygen in their atmospheres is present as CO_2, rather than O_2. Of course, as we have already discussed in Chapter 1, this unique and obviously important property of the earth's atmosphere is the direct result of photosynthesis by green plants. In addition to O_2, biological processes on the earth are responsible for the presence of many of the trace gases found in our atmosphere (see Table 3.4).

3.4.2. Physical Properties of the Atmosphere

The physical properties of a gas are characterized by three *state variables*: density (ρ), temperature (T), and pressure (p). The density (ρ) is simply the mass of the gas per unit volume and thus has units of grams per cubic centimeter. Temperature is a measure of the random kinetic energy contained in the individual mole-

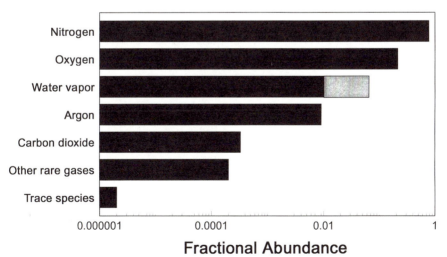

Fractional Abundance

Figure 3.9. The relative concentrations (by volume) or mole fractions of the major gases of the atmosphere. The shaded area represents the variable abundance of water vapor, typically from 1% to 5% near the earth's surface. Trace gases are defined as gases that are present in the atmosphere at concentrations of 1 part per million (i.e., 1 mole of gas for every million moles of other atmospheric gases) or less. A listing of some of the important trace gases of the atmosphere can be found in Table 3.4.

TABLE 3.4
Important C-, H-, N-, O-, P-, and S-Containing Trace Species of the Atmosphere

Species	Concentration (mole fraction)	Thermodynamically Equilibrated Concentration (mole fraction)[a]	Principal Sources
Methane (CH_4)	1.6×10^{-6}	10^{-145}	Biogenic
Carbon monoxide (CO)	$(0.5–2) \times 10^{-7}$	6×10^{-49}	Photochemical, anthropogenic
Ozone (O_3)	$10^{-8}–10^{-6}$	3×10^{-30}	Photochemical
Reactive nitrogen (NO_y)	$10^{-11}–10^{-6}$	10^{-9}	Biogenic, lightning, anthropogenic
Ammonia (NH_3)	$10^{-11}–10^{-9}$	2×10^{-60}	Biogenic
Particulate nitrate (NO_3^-)	$10^{-12}–10^{-8}$	—	Photochemical, anthropogenic
Particulate ammonium (NH_4^+)	$10^{-11}–10^{-8}$	—	Photochemical, anthropogenic
Nitrous oxide (N_2O)	3×10^{-7}	2×10^{-19}	Biogenic, anthropogenic
Hydrogen (H_2)	5×10^{-7}	2×10^{-42}	Biogenic, photochemical
Hydroxyl (OH)	$10^{-13}–10^{-11}$	5×10^{-28}	Photochemical
Peroxyl (HO_2)	$10^{-13}–10^{-11}$	4×10^{-28}	Photochemical
Hydrogen peroxide (H_2O_2)	$10^{-10}–10^{-8}$	1×10^{-24}	Photochemical
Formaldehyde (H_2CO)	$10^{-10}–10^{-9}$	9×10^{-96}	Photochemical
Sulfur dioxide (SO_2)	$10^{-11}–10^{-9}$	0	Anthropogenic, volcanic, photochemical
Dimethysulfide (CH_3SCH_3)	$10^{-11}–10^{-10}$	0	Biogenic
Carbon disulfide (CS_2)	$10^{-11}–10^{-10}$	0	Anthropogenic, biogenic
Carbonyl sulfide (OCS)	10^{-10}	0	Anthropogenic, biogenic
Particulate sulfate (SO_4^{2-})	$10^{-11}–10^{-8}$	—	Anthropogenic, photochemical
Particulate phosphate (PO_4^{3-})[b]	$10^{-12}–10^{-10}$	—	Terrestrial, anthropogenic

[a]Thermodynamically equilibrated concentrations for the gaseous species represent concentrations that would exist in equilibrium with a gas mixture containing 0.78 atm of N_2, 0.21 atm of O_2, 0.01 atm of H_2O, and 3.3×10^{-4} atm of CO_2 at a temperature of 298 K.
[b]There are no significant concentrations of phosphorus-containing trace gases in the atmosphere.
Source: Chameides W.L., and D. D. Davis, Chemistry in the troposphere, *Chemical and Engineering News,* **60,** 38–53, 1982.

 TEXT BOX 3.4   **ARGON: A "NOT-SO-RARE" RARE GAS**

Argon, along with the other elements that appear at the far right-hand side of the periodic table, is unreactive and almost always occurs in nature in monatomic form—that is, as a single atom. The stability of these elements is due to their electron configurations. Because they have completely filled outer electron shells, the rare gases have no propensity to gain, lose, or share electrons with other elements. As a result of this chemical stability, these elements are called the *inert* or *noble* gases. (Note that among its several definitions, the word "noble" is used to denote something that is chemically inert.) The paucity of the inert gases on the earth has also resulted in their being called the *rare* gases. A notable exception is Ar. As indicated in Figure 3.9, Ar is the fourth most abundant constituent of the atmosphere with a concentration approaching 1%. Can you explain why the rare gas Ar is not so "rare"? Here's a hint: It has something to do with the presence of a radioactive isotope of potassium (^{40}K) in the earth's crust.

cules of the gas; we will generally adopt Kelvin units for temperature where 0 K is absolute zero, 273 K is the freezing point of water, and 373 K is water's boiling point at sea level. Pressure is defined as the force per unit area exerted by a gas. This force is exerted by the myriad of collisions that result from the random motion of the individual molecules in the gas. We will use units of atmospheres to quantify pressure. (Recall from Chapter 2 that a gas at a pressure of 1 atm is defined as having an activity of 1.) One atmosphere is equal to 1.013×10^6 dyne cm^{-2} and is, not surprisingly, the average pressure of the earth's atmosphere at sea level.

The relationship between the three state variables is given by the ideal gas law:

$$p = \rho \, \frac{R^*}{m_{\text{atm}}} \, T = \rho R_{\text{atm}} T \tag{3.1}$$

where $R^* = 8.314 \times 10^7$ erg K^{-1} mole^{-1} = 0.082 atm K^{-1} (mole/liter)$^{-1}$ is the *universal gas constant*, $m_{\text{atm}} = 28.96$ g mole^{-1} is the mean gram-molecular weight of the atmosphere, and $R_{\text{atm}} = R^*/m_{\text{atm}} = 2.87 \times 10^6$ cm^2 s^{-2} K^{-1} is the gas constant for the atmosphere.

Atmospheric scientists use the vertical temperature structure of the atmosphere to divide it into four layers: the *troposphere*, the *stratosphere*, the *mesosphere*, and the *thermosphere*. As illustrated in Figure 3.10, the layers are characterized by al-

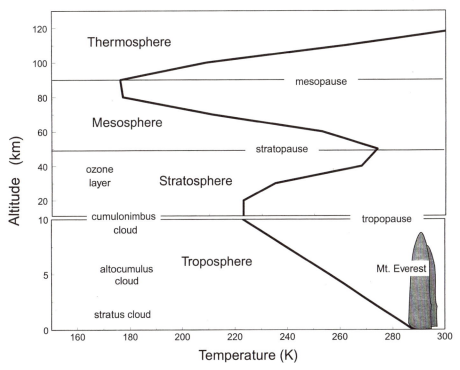

Figure 3.10. The average vertical variation of temperature as a function of altitude in the earth's atmosphere.

ternating positive and negative *lapse rates*, Γ, where the lapse rate is defined as the negative of the vertical temperature gradient; that is,

$$\Gamma = -\frac{dT}{dz} \tag{3.2}$$

where z is altitude. The three regions of the atmosphere with temperature maxima represent altitudes where solar photons are being preferentially absorbed and their energy is being converted to thermal energy. In the thermosphere, the photons being absorbed are primarily in the vacuum ultraviolet. In the stratosphere, the absorbed photons are mostly in the ultraviolet portion of the solar spectrum, and near the surface the photons in the visible portion of the solar spectrum are absorbed. Finally, note that the suffix "pause" is used to denote the nominal boundaries between the layers. Thus the *tropopause* is the boundary between the troposphere and stratosphere, the *stratopause* lies between the stratosphere and mesosphere, and the *mesopause* is the boundary between the mesosphere and thermosphere.

The vertical distribution of pressure and density in the atmosphere can be derived by considering the upward and downward forces on a parcel of air contained within a hypothetical volume of air having dimensions δx, δy, and δz (see Figure 3.11). If the volume is sufficiently small, we can assume that the density, ρ, of the air parcel is a constant and thus the total mass of the air parcel is simply $\rho \delta x \delta y \delta z$. There are two downward forces: (1) the force from the downward pull of gravity on the air parcel itself and (2) the force of pressure from the layer of gas immediately above the air parcel. The gravitational force is just the mass of the parcel times g, the acceleration due to gravity. The pressure force is the pressure at the top of the layer [i.e., $p(z + \delta z)$] times the area, in this case $\delta x \delta y$. Thus the total downward force is given by

$$F\downarrow = \rho g \delta x \delta y \delta z + p(z+\delta z)\delta x \delta y \tag{3.3}$$

Provided that δz is small, $p(z + \delta z)$ can be approximated in a Taylor's expansion:

$$p(z + \delta z) \cong p(z) + \frac{\partial p(z)}{\partial z} \delta z \tag{3.4}$$

Substituting in Equation (3.3), we obtain

$$F\downarrow = \left(\rho g + \frac{\partial p(z)}{\partial z}\right)\delta x \delta y \delta z + p(z)\delta x \delta y \tag{3.5}$$

The upward force on the parcel arises from the pressure exerted by gas immediately below the parcel, so that

$$F\uparrow = p(z)\delta x \delta y \tag{3.6}$$

Equating upward and downward forces, we obtain

$$\frac{\partial p(z)}{\partial z} = -\rho g \tag{3.7}$$

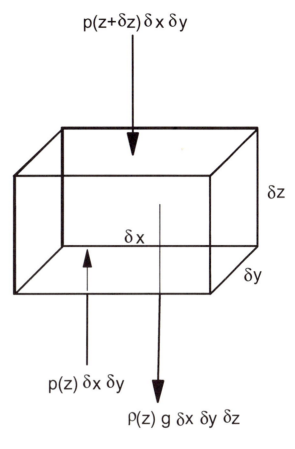

Figure 3.11. The upward and downward forces exerted on a hypothetical parcel of air having horizontal dimensions of δx and δy and a vertical dimension of δz. The downward forces consist of that exerted by gravity and the pressure exerted by the air residing directly above the parcel. The upward force arises from the pressure exerted by the air residing directly below the parcel. Note that the gravitational force is the mass times the acceleration due to gravity, g. Since the mass of the parcel is equal to the density, [rho], times the volume, the force of gravity on the parcel is given by [rho]$g\delta x\delta y\delta z$. Pressure, on the other hand, is a force per unit area. Thus, the pressure force on the parcel is given by the pressure at that altitude multiplied by the surface area over which the pressure is exerted—that is, $p\delta x\delta y$.

Equation (3.7) is known as the *hydrostatic equation* and states that, at equilibrium, atmospheric pressure decreases with height at the rate needed to produce a net upward buoyancy force that exactly compensates for the weight of the atmosphere above it. Combining Equations (3.1) and (3.7) and integrating, we obtain the barometric law which describes the vertical distribution of pressure in a hydrostatic atmosphere. For an atmosphere with a constant temperature, which turns out to be a reasonably good approximation for the earth, this integration yields

$$p(z) = p(0) \, \exp\!\left(\frac{-zg}{R_{atm}T}\right) = p(0) \, \exp\!\left(-\frac{z}{H}\right) \tag{3.8}$$

where $p(0)$ is the surface pressure and H ($=R_{atm}T/g$) is the atmospheric *scale height* and is the altitude at which the pressure (and density) of the atmosphere falls to $1/e$ of its surface pressure. For the earth, H is about 8.4 km. It follows therefore that well over two-thirds of the mass of the earth's atmosphere is within the troposphere. Note that from Equations (3.7) and (3.1), it can be shown that M_{TOT}, the total mass of any hydrostatic atmosphere, is given by

$$M_{TOT} = \rho(0)HA_{\bar{s}} \tag{3.9}$$

where A_s is the surface area of the planet. From this equation, we see that the scale height is simply the height the atmosphere would have if it were compressed down to a constant density atmosphere, with a density equal to its actual surface density. For the earth, $A_s = 5 \times 10^{18}$ cm^2 and $\rho(0) = 1.29$ g m^{-3}, and thus from Equation (3.9) we find that $M_{TOT} = 5 \times 10^{21}$ g, the same value indicated in Table 3.1.

3.4.3. Atmospheric Winds and Turbulent Mixing

Figure 3.12 is a schematic illustration of the average or climatological winds observed in the atmosphere, often referred to as the *general circulation*. From this figure, it can be seen that the atmosphere's circulation consists of a series of cells that circulate air latitudinally and vertically, coupled with a system of winds that move air alternately in the easterly and westerly directions.

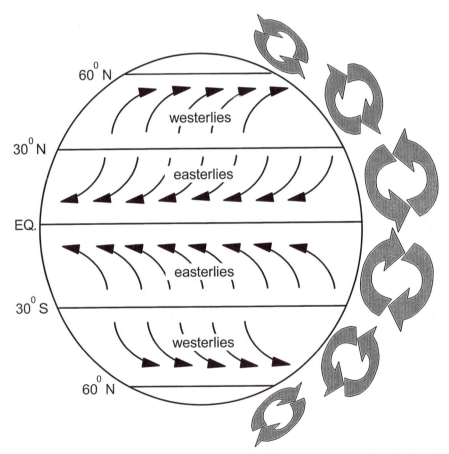

Figure 3.12. An idealized representation of the general circulation of the atmosphere. In the tropics, the winds are characterized by a thermal or Hadley cell circulation with rising winds at the equator, sinking winds at 30° latitude, and easterly winds at the surface. In the mid-latitudes, a reverse or Ferrel cell circulation occurs with sinking winds at 30°, rising winds at 60° latitude, and westerly winds at the surface.

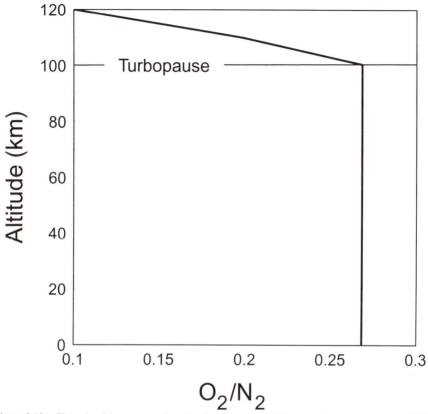

Figure 3.13. The ratio of the concentration of molecular oxygen (O_2) to that of molecular nitrogen (N_2) as a function of altitude. The turbopause, at an altitude of about 100 km, marks the point in the atmosphere where turbulence becomes too weak to keep the atmosphere well-mixed. Above the turbopause the constituents of the atmosphere begin to separate out by weight. This separation is evidenced by the precipitous drop in the ratio of the O_2 to N_2 above 100 km.

Although the average circulation of the atmosphere can be approximated by Figure 3.12, the atmospheric winds at any moment in time can differ significantly from this picture. As anyone can attest who has watched a flag waving about, winds can vary significantly over short periods of time. This high variability in the winds makes the atmosphere a turbulent environment where chemical constituents are rapidly mixed together. The effectiveness of this mixing process is indicated in Figure 3.13, where the observed ratio of the concentrations of molecular oxygen (O_2) to molecular (N_2) is plotted as a function of altitude. Because a molecule of O_2 is heavier than a molecule of N_2, we might expect this ratio to decrease as a function of altitude under the influence of gravity. However, in much the same way as shaking a bottle of salad dressing mixes the vinegar and oil in the bottle together, the mixing of the atmosphere by turbulent winds causes the ratio of O_2 to N_2 to remain essentially constant from the surface to about 100 km. Above 100 km (a level sometimes referred to as the *turbopause*), turbulent mixing becomes sufficiently light to allow the atmosphere to separate by weight; as a result, the ratio of O_2 to N_2 begins to decrease with height.

Because of the intensity of turbulent mixing in the atmosphere, the rate at which the atmosphere mixes is relatively short. Within a hemisphere, gases mix together in about 1–2 months. Mixing between the Northern Hemisphere and Southern Hemisphere is somewhat slower, typically requiring about 1–2 years. These mixing times are quite short compared to most other time scales of interest in biogeochemical cycles, and so we will usually be able to treat the atmosphere in our studies as a well-mixed reservoir of gases.

3.5. THE BIOSPHERE

In contrast to the other three spheres of the global biogeochemical system, the biosphere is not a specific, physical part of the earth, but rather the sum total of all living and dead organisms on the earth. Thus, the components of the biosphere are actually found in the lithosphere and hydrosphere,[1] and, to a very small extent, the atmosphere. The size or extent of the biosphere is most commonly expressed in terms of the total mass of organic carbon (C) on the earth. It is estimated that the biosphere currently contains a total of about 5×10^{18} g of C (or 5×10^6 Tg C). Roughly 20% of this C resides in living organisms while the rest resides in organic compounds produced by living organisms and arising from the decay of dead organisms. Most of this "dead" biospheric material is in the form of humic material, which is relatively resistant to chemical degradation in the environment. The residence time for living C in the biosphere is estimated to be about 10 years on the continents but only a month or two in the ocean. By comparison the residence time for "dead" C in the biosphere is on the order of a few to several decades.

The size of the reservoir of organic C in the biosphere and its rate of change can have a major effect on the fate of the carbon dioxide (CO_2) emitted to the atmosphere from the burning of fossil fuels and, by extension, a major effect on the impact of these emissions on climate via the *greenhouse effect*. For this reason, the determination of whether the biosphere is growing or shrinking and the rate at which it is growing or shrinking are subjects of intense scientific study. As it turns out, the answers to these questions can be found, at least in part, through an examination of the global biogeochemical cycles of C and O as we will find in Chapters 6 and 9.

3.5.1. Metabolic Processes

Living organisms can be distinguished from inanimate objects by the presence of *metabolic processes*—that is, processes involving chemical changes within the organism that allow it to continue to function and grow. These metabolic processes

---

[1]The biosphere occupies a surprisingly diverse array of environments in the earth system. For example, scientists have recently found evidence for bacteria living within sediments more than 500 m below the ocean floor (Parkes, R. J., et al., Deep bacterial biosphere in Pacific Ocean sediments, *Nature*, **371**, 410–413, 1994).

can be divided into two broad categories. *Anabolism*, or biosynthesis, is the construction of living protoplasm within the organism, and *catabolism* is the breaking down of organic matter to obtain energy to support the functions of the organism.

Organisms carry out biosynthesis in two ways: *Autotrophism* is the synthesis of living protoplasm from inorganic matter in the environment, and *heterotrophism* is the synthesis of protoplasm from other organic matter in the environment (i.e., through the ingestion of organic matter that had originally been produced by autotrophs). The process of biosynthesis requires energy. Autotrophs that obtain this energy from sunlight, such as green plants, are referred to as *photoautotrophs*, and autotrophs that obtain energy for biosynthesis from chemical reactions, like many bacteria, are referred to as *chemoautotrophs*. Virtually all heterotrophs derive the energy they require from catabolic reactions; thus, using the same classification scheme as that adopted for autotrophs, we would classify them as *chemoheterotrophs*. Examples of these three types of organisms and the chemical reactions these organisms use to carry out their metabolic processes are presented in Table 3.5. Note that our species, *Homo sapiens*, like all respiring animals, synthesizes protoplasm from organic matter (i.e., food) and obtains energy by oxidizing organic matter and, thus, is chemoheterotrophic.

TABLE 3.5
Metabolic Processes Found in Nature

Trophic Type	Example	Metabolic Process[a]	ΔG^0 (kJ/mole C)[b]
1. Chemoautotrophs[c]	Methane bacteria	$CO_2 + 4H_2 \rightarrow CH_4 + 2H_2O$	-114
2. Photoautotrophs[d]	Purple/green bacteria	$2CO_2 + H_2S + 2H_2O + h\nu \rightarrow$ $2\text{"}CH_2O\text{"} + H_2SO_4$	$+126$
	Green plants	$CO_2 + H_2O + h\nu \rightarrow$ $\text{"}CH_2O\text{"} + O$	$+478$
3. Chemoheterotrophs[e]	Fermenting bacteria	$2\text{"}CH_2O\text{"} \rightarrow CO_2 + CH_4$	-70
	Sulfate-reducing bacteria	$H_2SO_4 + 2\text{"}CH_2O\text{"} \rightarrow$ $2CO_2 + H_2S + 2H_2O$	-126
	Denitrifying bacteria	$4HNO_3 + 5\text{"}CH_2O\text{"} \rightarrow$ $2N_2 + 5CO_2 + 7H_2O$	-397
	Aerobes (or respirators)	$\text{"}CH_2O\text{"} + O_2 \rightarrow CO_2 + H_2O$	-478

[a]As discussed in Section 3.5.2, "CH_2O" is used in these stoichiometric equations as shorthand notation for the sugar glucose. Glucose has the chemical formula $C_6H_{12}O_6$, so that, stoichiometrically speaking, six "CH_2O" molecules are equivalent to one glucose molecule. The ΔG^0 values calculated for reactions involving CH_2O were actually determined by calculating the ΔG^0 for the equivalent reaction involving glucose and then dividing by six.
[b]The ΔG^0 values indicated here represent the change in Gibbs free energy for each mole of C involved in the reaction. The values were calculated using the data in the Appendix.
[c]Chemoautotrophs and chemoheterotrophs extract chemical energy from their metabolic processes, and thus the reactions for these processes have $\Delta G^0 < 0$.
[d]Photoautotrophs synthesize organic matter using energy from solar radiation. This process does not occur spontaneously, and thus the metabolic reactions for these processes have $\Delta G^0 > 0$.
[e]Respiration is the most thermodynamically favored of the metabolic pathways available to the chemoheterotrophs oxidizing organic matter. This very likely explains the dominant position of aerobes in the present-day, oxygen-rich biosphere. It is believed that chemoheterotrophs that use other metabolic pathways were probably more successful in earlier stages of the earth's evolution, when O_2 was either much less abundant or totally absent from the atmosphere [see Text Box 3.5]. Today, these other chemoheterotrophs are largely limited to anaerobic environments, where aerobes cannot exist.

| TEXT BOX 3.5 | A THUMBNAIL SKETCH OF THE EARTH'S HISTORY |

James Walker, in his book on *The Evolution of the Atmosphere*, describes the evolution of living systems on earth and the development of the biogeochemical cycles that support these systems as a series of opportunistic steps necessitated by the depletion of some critical nutrient in the biosphere. These steps are briefly outlined below.

4.6 billion years ago:
- Earth accretes.
- Atmosphere and ocean accumulate.
- Degassing from earth's interior leads to buildup of H_2.
- Abiotic processes (e.g., lightning) cause the buildup of organic C including CH_4.

4 billion years ago:
- Origin of life.
- Anaerobic fermenters (i.e., chemoheterotrophs) metabolize abiotically produced organic C.
- Biosphere limited by small supply of organic C.

3.5 billion years ago:
- Development of chemoautotrophy.
- Methane bacteria overcome dependence on abiotic-produced organic C by metabolizing CO_2 and H_2.
- Eventually population of chemoautotrophs limited by supply of H_2.

3 billion years ago:
- Development of photoautotrophy.
- Purple/green bacteria overcome dependence on H_2 by synthesizing organic C using solar energy and abiotically-produced H_2S as the reducing agent.
- Population of photosynthesizing bacteria limited by availability of H_2S.

2.7 billion years ago:
- Sulfate-reducing bacteria evolve.
- Bacteria utilize H_2SO_4 and organic C generated from purple/green bacteria and regenerate H_2S, thus creating a complete biogeochemical cycle.

2.5 billion years ago:
- Green plants evolve.
- Photosynthesis of organic C from H_2O and CO_2 energetically more efficient than metabolism used by purple/green bacteria.
- Photosynthesis causes a buildup of O_2, which is a strong oxidant and a threat to anaerobes and eventually all organisms if concentration becomes too high.

1.8 billion years ago:
- Aerobic respiration evolves.
- Aerobes utilize O_2 and organic C from green plants

TEXT BOX 3.5	A THUMBNAIL SKETCH OF THE EARTH'S HISTORY (*CONT.*)

	and regenerate CO_2, creating energetically efficient biogeochemical cycle.
Present:	• Human population approaches 6 billion.
	• Use of fossil fuels and destruction of biomass leads to dramatic increases in atmospheric CO_2 and other greenhouse gases.
	• Production of synthetic halogenated chemicals threaten stratospheric ozone layer.
Future:	• ?????

TABLE 3.6
The Percent (by weight) of Chemical Compounds in Various Organisms

Compound Type	Organism			
	Algae	Bacteria	Copepods	Terrestrial Plants
Carbohydrates	30	40	2	45
Proteins	40	50	75	5
Lipids	5	10	15	1
Lignin	0	0	0	20
Minerals (i.e., inorganic)	25	0	8	29

3.5.2. The Composition of the Biosphere

As indicated in Tables 3.6 and 3.7, the organisms and organic wastes that make up the biosphere are composed of a complex mixture of *organic polymers* (i.e., organic molecules produced by combining two or more simpler organic molecules). These organic polymers include carbohydrates (sugars and starches), proteins, and lipids, as well as lignin in woody terrestrial plants. These compounds, in turn, contain varying amounts of C, H, O, and N (and a host of other trace elements). In order to make the treatment of biospheric processes in global biogeochemical cycles tractable, it is desirable to identify a single chemical formula that represents the average composition of all biospheric organisms in total. Thus far in our discussions, we have been using "CH_2O" to do just that. In doing so, we have been making the implicit assumption that organic matter in the biosphere is composed of the elements C, H, and O and that these elements appear in a stoichiometric ratio of 1:2:1.

A C:H:O ratio of 1:2:1 turns out to be the same ratio as that found in the sugar glucose, which has a chemical formula of $C_6H_{12}O_6$. (Note that by combining six "CH_2O" molecules we obtain the stoichiometric equivalent to $C_6H_{12}O_6$.) So, by using "CH_2O" to represent the stuff of organisms, we have been treating the biosphere as if it were composed exclusively of glucose. There are, in fact, good reasons for treating the biosphere as the stoichiometric equivalent to glucose. Glucose is the

TABLE 3.7
Average Relative Amounts of C, H, O, and N Found in Organic
Polymers in Nature

Organic Polymer	Elemental Ratios		
	H:C	O:C	N:C
Carbohydrates	1.67	0.83	0
Proteins	1.54	0.38	0.27
Lipids	2	0.1	0
Lignin	1.1	0.37	0
Kerogen	0.99	0.1	0.02
Coal	0.76	0.11	0
Humus	1.2	0.64	0.03
Redfield Biomass	2.48	1.04	0.15

simplest of the carbohydrates found in nature and is the building block from which other molecules of the biosphere are polymerized or synthesized. For instance, glucose is the initial organic product formed by green plants in photosynthesis. In most chemoheterotrophs, glucose is the end product of the digestive process that allows for the assimilation of more complex carbohydrates.

Despite glucose's fundamental role in the metabolic processes of both autotrophs and chemoheterotrophs, the use of "CH_2O" to represent all the biosphere presents some problems. The most serious of these is the fact that the living organisms that make up the biosphere are not just composed of C, H, and O. They contain other nutrient elements; for our purposes here, the most notable of these other elements are N and P. If we are to study the interaction of these other nutrient elements with the biosphere, we must be able to quantify the rates at which these elements are incorporated into and lost from the biosphere. Moreover, because organisms on land tend to be fundamentally different in composition from those in the ocean, we must do this for both the terrestrial and oceanic biosphere. Let's turn to the oceanic biosphere first.

3.5.2.1. Living Organisms In The Ocean

The rates at which N and P are incorporated into and lost from the biosphere are most easily inferred by relating these rates to the rates at which C is incorporated and lost from the biosphere. For organisms in the ocean, this is accomplished by invoking the so-called *Redfield ratio*. The Redfield ratio—named after Alfred Redfield, whose measurements in the late 1950s and early 1960s first characterized the elemental composition of oceanic phytoplankton—specifies the average relative amounts of C, N, and P in oceanic organisms. Initially estimated by Redfield to be 80:15:1, the Redfield ratio for C:N:P in oceanic biota is currently believed to be about 106:16:1.

The Redfield ratio can be integrated into our overall metabolic scheme for the formation and breakdown of living matter in the ocean by including appropriate amounts of ammonia (NH_3) and phosphoric acid (H_3PO_4) in the relevant stoichiometric reactions used to represent these metabolic processes. For instance, in our initial simple representation using "CH_2O" we used

$$CO_2 + H_2O + h\nu \rightarrow CH_2O + O_2 \qquad ((R1.2)$$

to represent photosynthesis and

$$CH_2O + O_2 \rightarrow CO_2 + H_2O \qquad (R1.3)$$

to represent respiration and decay of organic matter. To include N and P, we can represent oceanic photosynthesis by

$$106CO_2 + 106H_2O + 16NH_3 + H_3PO_4 + h\nu$$
$$\rightarrow (CH_2O)_{106}(NH_3)_{16}(H_3PO_4) + 106O_2 \qquad (R3.1)$$

or, equivalently,

$$106CO_2 + 106H_2O + 16NH_3 + H_3PO_4 + h\nu$$
$$\rightarrow C_{106}H_{263}O_{110}N_{16}P + 106O_2 \qquad (R3.1)$$

Similarly, inclusion of N and P in the process of respiration would yield the opposite of (R3.1)—that is,

$$C_{106}H_{263}O_{110}N_{16}P + 106O_2 \rightarrow 106CO_2 + 106H_2O + 16NH_3 + H_3PO_4 \qquad (R3.2)$$

In this revised formulation, "$C_{106}H_{263}O_{110}N_{16}P$" instead of "$CH_2O$" is used to represent an average molecule of oceanic, biospheric matter. Note that in this new representation the biospheric material contains C, N, and P in the Redfield ratio of 106:16:1, but is relatively richer in O and H than in CH_2O because of the extra atoms of these elements supplied from NH_3 and H_3PO_4.

While "$C_{106}H_{263}O_{110}N_{16}P$" is a more comprehensive representation of the ocean biosphere, it also is not without shortcomings. For one, it does not account for the presence of other trace elements (e.g., iron or silica) in living organisms. If we were to attempt to develop a biogeochemical cycle for one of these other elements, we would have to first determine this element's relative abundance in the biosphere as we did for N and P using the Redfield ratio. Another shortcoming has to do with the relative biospheric abundances of H and O implied by "$C_{106}H_{263}O_{110}N_{16}P$." Note that this formulation yields a H:C ratio of almost 2.5 and a O:C ratio of a little more than 1. However, as can be seen in Table 3.7, the organic polymers found in oceanic organisms typically contain significantly less H and O. (The low levels of H and O in the organic polymers listed in Table 3.7 might at first seem somewhat surprising. Since glucose contains C, H, and O in a ratio of 1:2:1 and virtually all organic polymers in nature are generated from glucose, one might be inclined to conclude that the C:H:O ratio in these organic polymers must also be 1:2:1. However, during the polymerization process by which complex organic molecules are generated from glucose, water is released. The release of this water results in the lower abundances of H and O indicated in Table 3.7.)

To more accurately reflect the lower abundances of H and O in the organic polymers found in oceanic organisms, we must make another revision to our biospheric molecule. This can be accomplished by assuming that biomass is synthesized (for instance via photosynthesis) by combining CO_2 and H_2O in a ratio greater than 1:1. For instance, when choosing a ratio of 106:64 we obtain a slightly different stoichiometric representation for photosynthesis and respiration:

$$106CO_2 + 64H_2O + 16NH_3 + H_3PO_4 + h\nu$$
$$\rightleftarrows C_{106}H_{179}O_{68}N_{16}P + 106O_2 \qquad (R3.3)$$

This yields a formula for oceanic biomass—"$C_{106}H_{179}O_{68}N_{16}P$"—that is consistent with the Redfield ratio for C:N:P, while at the same time having H:C and O:C ratios that are closer to those actually found in the organic polymers that make up the biosphere. And this is the formula we will use to represent the living portion of the oceanic biosphere.

3.5.2.2. Living Organisms On The Land

Because the mass of the terrestrial biosphere is dominated by trees which are relatively rich in lignin, the terrestrial biomass is characterized by significantly lower O:C, H:C, N:C, and P:Cc ratios than oceanic biomass (see Tables 3.6 and 3.7). We therefore choose a stoichiometric representation for terrestrial biomass— "$C_{830}H_{1230}O_{604}N_9P$"—and a stoichiometric reaction for photosynthesis and respiration and decay by which this biomass is produced and destroyed that reflects these lower ratios:

$$830CO_2 + 600H_2O + 9NH_3 + H_3PO_4 + h\nu$$
$$\rightleftarrows C_{830}H_{1230}O_{604}N_9P + 830O_2 \qquad (R3.4)$$

3.5.3. Primary Production

The rate at which C is assimilated or fixed by green plants in photosynthesis is referred to as the rate of *primary production*. Because so many of the earth's biogeochemical cycles are driven by the rate at which green plants produce organic material via photosynthesis, the primary production rate represents one of the fundamental parameters of the earth system and one that we will be turning to numerous times in our analyses of these cycles.

In defining primary production, it is useful to distinguish between the *gross primary production rate*, which denotes the total rate at which C is assimilated by green plants, and the *net primary production rate*, which denotes the C assimilation rate after subtracting off the rate of C loss due to respiration. In other words,

$$NPP = GPP - R_{gp} \qquad (3.10)$$

where NPP is the net primary production rate, GPP is the gross primary production rate, and R_{gp} is the green-plant respiration rate.

For the terrestrial biosphere, it is estimated that about 50% of the gross primary production remains in the biosphere as net primary production on a global basis, with GPP equal to about 1×10^5 tg C year^{-1} and NPP about 5×10^4 tg C year^{-1}. Initially, this NPP input of C to the terrestrial biosphere is stored in the living biomass, where it may be ingested and utilized by heterotrophs. While some portion of this C is lost from the biosphere via heterotrophic respiration, the vast majority of it is transferred to the dead terrestrial biosphere as litter fall, incorporated into humic material and finally transferred back to the atmosphere as CO_2 upon decomposition. In the absence of anthropogenic perturbations, the size of the terres-

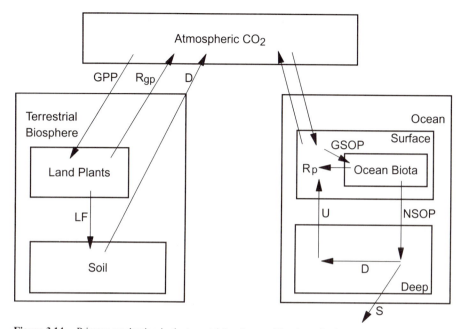

Figure 3.14. Primary production in the terrestrial and ocean biosphere. In the terrestrial biosphere, net primary production (NPP) is defined as the difference between gross primary production (GPP) and green-plant respiration (R_{gp}). For an unperturbed and balanced terrestrial biosphere, the input of C from NPP is balanced by loss to the soils, largely in litter fall (LF), and the subsequent decay (D) of this organic matter. In the ocean biosphere, we will use net surface ocean productivity (NSOP) to represent the fraction of the gross surface ocean productivity (GSOP) that escapes respiration in the photic zone (R_p) and sinks into the deep ocean. Most of the NSOP is eventually returned to the surface ocean via decay (D) and upwelling (U). However, a small fraction is lost from the ocean via sedimentation (S). Recall that the organic C lost from the system via sedimentation is ultimately returned to the atmosphere in the rock cycle as a result of uplift and weathering of the sedimentary material.

trial biosphere would be in steady state as long as the rate of litter fall (LF) and decomposition (D) are equal to NPP (see Figure 3.14). Whether this is in fact the case, as well as the extent to which humankind has upset this balance, will be discussed in our analysis of the global C cycle in Chapter 6.

When considering the ocean biosphere, the concepts of gross and net primary production are less useful. Because of the short lifetime of phytoplankton and the resulting rapid recycling of organic C that takes place within the surface ocean, it is far more useful to distinguish between the C that is assimilated into phytoplankton residing in the surface ocean, which we will refer to as the *gross surface ocean production rate* (GSOP), and the organic C that sinks into the deep ocean before decomposing, which we will refer to as *net surface ocean production rate* (NSOP). The relationship between the rates of GSOP and NSOP and phytoplankton respiration (R_p), deep ocean organic C decay (D), and upwelling (U) are schematically illustrated in Figure 3.14. The importance of NSOP to biogeochemical cycles arises from the fact that it sequesters C and related nutrient elements into the deep ocean where they cannot participate in further cycling through the biosphere until upwelling returns them to the surface ocean. Recall from our earlier discussion of the nutrient cycle that the upwelling time is long, on the order of 1000 years (see Figure 3.5).

In contrast to the terrestrial biosphere where NPP is about one-half of GPP, the ratio of net surface ocean productivity to gross ocean productivity is about 1 to 10. On a global basis, GSOP is estimated at about 4×10^4 tg C year^{-1}, while NSOP is estimated at about 0.4×10^4 tg C year^{-1}.

3.6. CONCLUSION

The four spheres of the earth system—lithosphere, hydrosphere, atmosphere, and biosphere—will prove to be a useful framework for cataloging the major reservoirs where the elements are sequestered and for categorizing the kinds of biogeochemical processes that occur on the earth. To facilitate easy reference to the critical quantities that characterize these reservoirs and the flow of material into and out of them, we have summarized the masses, fluxes, and lifetimes of the spheres in Table 3.8 and schematically illustrated the key processes and the interactions of the spheres in Figure 3.15. Suffice it to say that we will be returning to this table and figure often in our subsequent discussions.

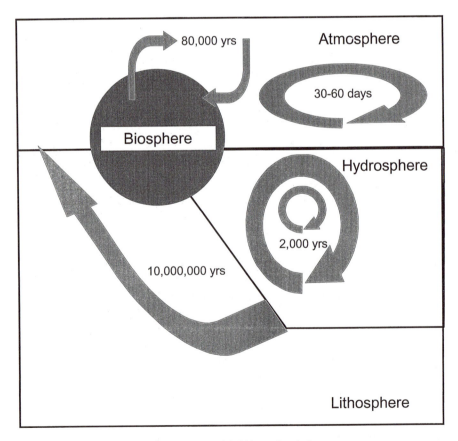

Figure 3.15. Characteristic time scales in the global biogeochemical system.

TABLE 3.8
Summary of the Earth System's Current Metabolic State: Reservoir Masses, Fluxes, and Cycling Times

Sphere	Parameter	Notes
Hydrosphere	Reservoir masses (Tg) Surface ocean, 1.1×10^{11} Deep ocean, 1.3×10^{12} Total ocean, 1.4×10^{12}	Reservoir masses estimated assuming ocean surface area = 3.5×10^{18} cm^2, surface ocean depth = 300 m, and deep ocean depth = 3700 m.
	Fluxes (Tg year^{-1})[a] River runoff, 4.0×10^7 Precipitation, 3.9×10^8 Evaporation, 4.3×10^8 Surface/deep ocean exchange, 7.3×10^8 Cycling times (years) Surface ocean, ~ 100 Deep ocean, ~1000	River runoff, precipitation, and evaporation rates taken from reference 1.[b] Surface/deep ocean exchange rate calculated assuming an exchange coefficient of 2 m/year (reference 3).
Lithosphere	Reservoir masses (Tg) Continental crust, 2×10^{13} Soils, 2×10^8 Oceanic crust, 7×10^{12} Sediments, 2×10^{12}	Soil and ocean sediment reservoirs estimated assuming: lithospheric density = 2.5 g m^{-3}, depth of sediments = 2 km, and depth of soils = 60 cm.
	Fluxes (Tg year^{-1}) Continental weathering rate, 2×10^4 Sedimentation rate, 2×10^4	Weathering rate estimated from mass of dissolved and particulate matter carried in rivers to ocean (reference 1). Sedimentation rate assumed to be equal to weathering rate.
	Cycling times (years) Soils, ~10^4 Ocean sediments, ~10^8 Continental crust, ~10^9	
Atmosphere	Reservoir masses (Tg) Atmosphere, 5×10^9 Troposphere, 4.5×10^9 Mixing times (years) Intrahemispheric, ~0.1 Interhemispheric, ~2	
Biosphere	Reservoir masses (Tg C) Living Terrestrial biosphere, 8.3×10^5 Ocean biosphere, 1.8×10^3 Dead (humus) Terrestrial, 1.5×10^6 Ocean, 3×10^6 Total biosphere, 5×10^6	Estimates taken from references 1, 2, and 4.
	Primary Production (Tg C year^{-1}) Terrestrial biosphere Gross primary production, 9.6×10^4 Net primary production, 4.8×10^4 Ocean biosphere Gross surface ocean production, 4×10^4 Net surface ocean production, 0.4×10^4	Estimates taken from references 1, 2, and 4.

TABLE 3.8 (*Continued*)

Sphere	Parameter	Notes
Biosphere (continued)	Cycling Times (years) 　Living 　　Terrestrial biosphere, ~10 　　Ocean biosphere, ~0.1 　Dead (humus) 　　Terrestrial, ~30 　　Ocean, ~75 　C:H:O:N:P ratios 　　Terrestrial biosphere, 　　830:1230:604:9:1 　　Ocean biosphere, 　　106:179:68:16:1	 See text.

[a]1 Tg $= 1 \times 10^6$ metric tons $= 1 \times 10^{12}$ g.
[b]References: (1) Schlesinger, W. H. *Biogeochemistry—An Analysis of Global Change*, Academic Press, New York, 1991. (2) Bolin, B. The Carbon Cycle, in *The Biosphere*, Scientific American Book, W. H. Freeman and Co., San Francisco, 1970. (3) Broecker, W. S., *Quartenary Research*, **1**, 188, 1971. (4) Whittaker, R. H., and G. E. Likens, Carbon in the biota, in *Carbon and the Biosphere*, CONF 720510, National Technical Information Service, Washington, D.C., 1973.

SUGGESTED READING

Chameides, W. L., and D. D. Davis, Global tropospheric chemistry, *Chemical and Engineering News*, **60**, 38–52, 1982.

Hutchinson, G. E., The biosphere, *Scientific American*, September 1970.

Kasting, J. F., O. B. Toon, and J. B. Pollack, How climate evolved on the terrestrial planets, *Scientific American*, 90–97, February 1988.

Schlesinger, W. H., *Biogeochemistry: An Analysis of Global Change*, Academic Press, New York, 1991.

Swanson, C. P., *The Cell*, Prentice-Hall, Englewood Cliffs, NJ, 1969.

Turekian, K. K., *Oceans*, Prentice-Hall, Englewood Cliffs, NJ, 1976.

Walker, J. C. G., *Evolution of the Atmosphere*, Macmillan, New York, 1977.

PROBLEMS

1. Using a topographic map of the earth, identify regions on the globe where one is apt to find divergent plate boundaries and convergent plate boundaries.

2. Use the data in Figure 3.4 and Table 3.2 to estimate the total number of grams and number of moles of P and N in the ocean.

3. Show that Equation (3.9) can be derived from Equations (3.7) and (3.1) for any distribution of atmospheric temperatures.

4. Use the data in Figure 3.9 and Equation (3.9) to estimate the total number of grams and number of moles of N and O in the earth's atmosphere.

5. Use the data in Table 3.8 to estimate the total number of grams and number of moles of P and N in the terrestrial and oceanic biospheres.

6. Use the data in Table 3.8 to estimate the rate at which N and P is assimilated by the terrestrial and oceanic biospheres as a result of photosynthesis.

The Mathematics of Simulating Biogeochemical Cycles

4

> "The earth, whether viewed globally or on a local scale, is certainly a dynamic and evolving chemical system. . . . If a geochemist is interested in describing the nature of the distribution of the elements in the earth, it is only appropriate to do so within a general . . . (mathematical) . . . framework."
>
> A. C. Lasaga, Dynamic treatment of geochemical cycles: Global kinetics, in *Kinetics of Geochemical Processes, Reviews of Mineralogy,* **8,** 69–109, 1988.

4.1. INTRODUCTION

One of the primary goals of the study of global biogeochemical cycles is to be able to predict how a perturbation in one part of a cycle will affect the rest of the cycle. For example, since the Industrial Revolution, humankind has been mining fossil fuels stored in the lithosphere and burning them. Since fossil-fuel burning converts organic carbon to CO_2 gas, it is fairly obvious that this process will cause an increase in atmospheric CO_2. The problem is that we need more quantitative information. Because increasing concentrations of atmospheric carbon dioxide (CO_2) could cause a global warming through an enhanced "greenhouse effect," we need to know how large the CO_2 increase will be and how long the increase will persist. The answers to these questions are not so obvious. To answer them we need a mathematical description of the cycle and the various biogeochemical processes that cause C to flow from one sphere of the earth system to another.

In this chapter we discuss how one goes about developing this mathematical description of a biogeochemical cycle. We begin with a review of the differential equations that govern biogeochemical cycles and their solutions and we then introduce *BOXES*, the computer software supplied with the text for the reader to carry out these simulations independently.

4.2. THE LINEAR BOX MODEL

To keep things relatively simple, we will adopt the "linear box model" approximation. In this approximation, a biogeochemical cycle is treated as a coupled system of reservoirs or boxes. The individual components of the earth system (e.g., the atmosphere, hydrosphere, lithosphere, and biosphere) represent the reservoirs in the system. Within each reservoir resides a time-varying amount of the specific element whose biogeochemical cycle we are trying to model. Biogeochemical processes that affect the element are treated as pathways by which the elements are transferred from one reservoir to another.

The word *linear* is used because the rates of transfer from one box to another are assumed to be linearly proportional to the amount of the element in each reservoir. For instance, in the case of the biogeochemical cycle of C, photosynthesis would be represented in this approximation as a transfer of carbon from the atmospheric reservoir (or box) to the biospheric reservoir. The rate of this transfer would be assumed to be proportional to the abundance of atmospheric CO_2.

Our description of this approach and the relevant mathematics needed to simulate biogeochemical cycles using this approach begins with a simple illustrative example involving a fictitious university—the University of Biogeochemistry.

4.3. SIMPLE EXAMPLE: THE UNIVERSITY OF BIOGEOCHEMISTRY–WORLD CYCLE

Imagine a fictitious world that consists of only two types of people: those people who do not attend the University of Biogeochemistry and those people who attend the University of Biogeochemistry. As illustrated in Figure 4.1, we can use a simple two-box model to represent this cycle and the flow of people (or students) into the university (as freshman) and out of the university upon graduation. We let

C_1 = number of people in the world who do not attend the university

C_2 = number of people (or students) attending the university

$F_{1 \to 2}$ = rate of flow of people from the world to the university (i.e., the admission rate)

$F_{2 \to 1}$ = rate of flow of people from the university to the world (i.e., the graduation rate)

In our simple (perhaps ideal) model world, no one is allowed to flunk out of the university and no one is allowed to die.

We now make the linear box model approximation. We assume that the flow of people out of any box is proportional to the population of the people in the box. Thus

$$F_{1 \to 2} = k_{1 \to 2} C_1$$
$$F_{2 \to 1} = k_{2 \to 1} C_2$$

(4.3.1)

where $k_{i \to j}$ is the proportionality or transfer constant for the flux of people from the ith to the jth reservoir and has units of inverse time (e.g., s^{-1}).

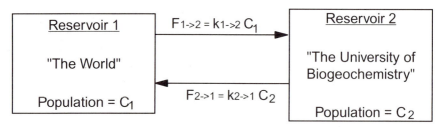

Figure 4.1. Schematic representation of the fictitious cycle of people between the world and the University of Biogeochemistry. C_1 and C_2 are used to denote the populations of the world and the university, respectively. The flow of people from the world to the University is represented by $F_{1 \to 2} = k_{1 \to 2}C_1$ and the flow of people from the university to the world is represented by $F_{2 \to 1} = k_{2 \to 1}C_2$, where $k_{1 \to 2}$ and $k_{2 \to 1}$ are the so-called *transfer coefficients*. Our initial problem will be to estimate reasonable values for the C's and k's. We will then use these values to construct a mathematical description of the cycle and use this description to analyze how the cycle evolved over it time and how it might respond to changing conditions.

Now we turn to the real work—determining the C's, F's, and k's of the cycle. We start with C_1, the number of people in the world.

4.3.1. C_1, the World Population

The world population is, of course, a quantity that is fairly well known (about 5.4 billion) and we could very easily look it up in an almanac. But it is instructive to try to come up with an estimate on our own since we will have to make these sort of estimations when we consider actual biogeochemical cycles. So, instead of looking up the world population, let's try to estimate it on our own. Any ideas?

One of the common methods biogeochemists use to estimate reservoir amounts is through sampling—that is, sampling the population in a portion of the reservoir and extrapolating to the entire reservoir. Consider, for instance, Fulton County, where the authors' university is located. Fulton County has a population of about 1 million people and covers about 66 km^2. If we assume that Fulton County's population density is representative of the globe, it would follow that

$$C_1 = (1 \times 10^6 \text{ people}) \times \left(\frac{5 \times 10^8 \text{ km}^2}{66 \text{ km}^2} \right)$$

$$\approx 10^4 \text{ billion people}$$

(4.3.2)

This is, of course, a gross overestimate of the world population. The fact is that Fulton County lies within a major metropolitan area (Atlanta, Georgia) and has a population density well above the global average. There is an important lesson here that we must always bear in mind when we are developing actual biogeochemical cycles: If we base our estimates on extrapolations from unrepresentative samples, we may incur very large errors. In general, we will increase the likelihood of obtaining accurate estimates of global populations and rates by sampling several different subpopulations and regions.

A more accurate estimate for C_1 can be obtained by accounting for the facts that continents only comprise 30% of the earth's surface, cities only comprise about

0.1% of the total land mass of the earth, and roughly 40% of the world's population currently reside in urban areas. With these additional statistics, we estimate that

$$C_1 = (1 \times 10^6 \text{ people}) \times \left(\frac{5 \times 10^8 \text{ km}^2}{66 \text{ km}^2} \right) \times \frac{0.3 \times 0.001}{0.4}$$

$$\approx 5.5 \text{ billion people} \tag{4.3.3}$$

This is close to the actual present-day value for C_1 and is the value we will adopt in our calculations here.

4.3.2. C_2, The University Population

Since the University of Biogeochemistry is a fictitious university, we are free to choose any number for its population. For simplicity, the value we will use for C_2 in the text will be based on the population of the authors' university. (Readers are encouraged to use their own university or any other institution to create their own cycle.) There are a variety of approaches that could be taken for estimating the population of a university. One such approach developed in conjunction with the students who have taken our course is adopted below.

Based on the authors' and our students' experience, we estimate that the average class on our campus contains about 20 students and that, at any given time, there are about 100 classes ongoing on our campus. Moreover, we have eight lecture periods per day and our students take, on average, about 1.5 classes per day. This suggests that about 20% of the students in our university are attending class during any given lecture period. Thus,

$$C_2 \approx \left(\frac{20 \text{ students}}{\text{class}} \right) \times (100 \text{ classes}) \div \left(0.2 \ \frac{\dfrac{\text{students}}{\text{class}}}{\dfrac{\text{students}}{\text{university}}} \right)$$

$$\approx 10{,}000 \text{ students} \tag{4.3.4}$$

This turns out to be a fairly good estimate of the population of our university and is the value we will use here.

4.3.3. $k_{2 \to 1}$, Transfer Coefficient from University to World

The transfer coefficient $k_{2 \to 1}$ is the easiest of the two transfer coefficients in our cycle to estimate since we know something about the flow of students out of universities. So let's consider it first.

The average time that a student resides at a university is generally 4 years. We thus define, $\tau_{2 \to 1} = 4$ years as the *residence time* for a student at the University of Biogeochemistry before returning to the world. The flux of material from a reservoir is defined as its population divided by its residence time, and thus

$$F_{2 \to 1} = \frac{C_2}{\tau_{2 \to 1}} \tag{4.3.5}$$

Combining Equation (4.3.1) with (4.3.5) gives

$$k_{2\to 1} = \frac{1}{\tau_{2\to 1}} = \frac{1}{4 \text{ years}} = (0.25) \text{ years}^{-1} \tag{4.3.6}$$

4.3.4. $k_{1\to 2}$, Transfer Coefficient from World to University

Determination of the transfer coefficient from the world to the university is more difficult because we have very little information on this type of flow. In fact, an inability to estimate the value of a transfer coefficient is a problem we will often face when constructing biogeochemical cycles. It behooves us to find some way around this problem. One solution is to adopt the *steady-state assumption*—that is to assume that the system is currently in equilibrium. At equilibrium, there is no net flow into or out of any reservoir. Thus, at steady state

$$F_{1\to 2} = F_{2\to 1} \tag{4.3.7}$$

Using Equations (4.3.1) and (4.3.7) and substituting for the values defined above for C_1, C_2, and $k_{2\to 1}$, we now have enough information to derive a value for $k_{1\to 2}$:

$$k_{1\to 2} = \frac{C_2 k_{2\to 1}}{C_1} = \frac{(10,000 \text{ people}) \times (0.25 \text{ years}^{-1})}{5.5 \times 10^9 \text{ people}}$$

$$= 4.6 \times 10^{-7} \text{ years}^{-1} \tag{4.3.8}$$

Equation (4.3.8) has an interesting implication. Recall that the residence time is equal to the inverse of the transfer coefficient. Therefore, it follows that $\tau_{1\to 2}$, the residence time for people to remain in the world before going to the university, is given by

$$\tau_{1\to 2} = \frac{1}{k_{1\to 2}} \approx 2 \text{ million years} \tag{4.3.9}$$

A residence time of 2 million years is a rather long time, especially given the average life expectancy of human beings today of some 70 years. What did we do wrong?

The answer is that we left an important set of processes out of our cycle, namely, birth and death. Without birth and death, there is no net loss or gain in the total number of people in the system; in other words,

$$C_1 + C_2 = 5,500,010,000 \text{ people} = \text{constant} \tag{4.3.10}$$

Thus, the total number of people remains fixed (a characteristic of all closed cycles) and they simply cycle back and forth from the world to the university. Because there are so many more people in the world than at the university in this cycle, people have to stay in the world a very long time (i.e., 2 million years) before they get a 4-year stint at the university. Another way of interpreting our result is to note that if the average life span of a human being is 70 years and the residence time to go to the university is 2 million years, then the chances of any individual going to the University of Biogeochemistry is approximately 70/2,000,000, or about

1 in 30,000. In other words, those of us who get to learn about biogeochemical cycles are actually rather lucky.

4.4. USING DIFFERENTIAL EQUATIONS TO SIMULATE THE UNIVERSITY–WORLD CYCLE

In Section 4.3 we set up the steady-state cycle of people between our fictitious University of Biogeochemistry and the world. However, the formulation we developed does not enable us to analyze how this cycle might have evolved over time or how it might respond to a perturbation. In this section we will illustrate how differential equations can be used to study our fictitious university–world cycle in more detail.

Since our cycle consists of two boxes, we will need two differential equations to represent the time rate of change of the populations of the two reservoirs or boxes. From Figure 4.1 it should be relatively obvious that the differential equations are given by

$$\frac{dC_1(t)}{dt} = F_{2\to1} - F_{1\to2} = k_{2\to1}C_2(t) - k_{1\to2}C_1(t)$$

$$\frac{dC_2(t)}{dt} = F_{1\to2} - F_{2\to1} = k_{1\to2}C_1(t) - k_{2\to1}C_2(t)$$

(4.4.1)

where t is time (in years) and we now use $C_i(t)$ to represent the population of each reservoir. (We write the C_i's as functions of the variable t to explicitly indicate the fact that the populations are functions of time.)

An important characteristic of this, and any other properly formulated closed cycle, is that the sum of all the time derivatives is zero; that is,

$$\frac{dC_1(t)}{dt} + \frac{dC_2(t)}{dt} = 0$$

(4.4.2)

This ensures that there is no production or loss of the material being cycled. This characteristic provides an easy method for checking for mistakes whenever you write down the differential equations for a closed cycle; if the sum of all the time derivatives does not add to zero, you have made a mistake in one or more of the equations.

Equation (4.4.1) represents a coupled set of linear, homogeneous differential equations. The solution to this type of system of differential equations can be obtained by assuming an e^{wt} dependence on time; that is,

$$C_1(t) = \Phi_1 e^{Et}$$

$$C_2(t) = \Phi_2 e^{Et}$$

(4.4.3)

where E, Φ_1, and Φ_2 are constants that we must determine. Substitution of Equations (4.4.3) into Equations (4.4.1) yields

$$\Phi_1 E e^{Et} = k_{2\to1}\Phi_2 e^{Et} - k_{1\to2}\Phi_1 e^{Et}$$

$$\Phi_2 E e^{Et} = k_{1\to2}\Phi_1 e^{Et} - k_{2\to1}\Phi_2 e^{Et}$$

(4.4.4)

Now by dividing Equations (4.4.4) by e^{Et} and rearranging terms, we obtain

$$0 = -(E + k_{1\to2})\Phi_1 + k_{2\to1}\Phi_2$$
$$0 = k_{1\to2}\Phi_1 - (E + k_{2\to1})\Phi_2$$

(4.4.5)

Thus, through a series of simple manipulations we have essentially converted our problem from one of integrating a system of differential equations to one of solving a system of algebraic equations. Students of linear algebra will recognize that Equations (4.4.5) are equivalent to a vector-matrix equation of the form

$$(M)\vec{\Phi} = 0 \qquad (4.4.6)$$

where **M** is a 2×2 matrix defined by

$$\mathbf{M} = \begin{matrix} -E - k_{1\to2} & k_{2\to1} \\ k_{1\to2} & -E - k_{2\to1} \end{matrix} \qquad (4.4.7)$$

and Φ is a 2-dimensional vector given by

$$\Phi = \begin{matrix} \Phi_1 \\ \Phi_2 \end{matrix} \qquad (4.4.8)$$

A non-trivial solution to Equation (4.4.6), that is, a solution where $\Phi \neq 0$, is one where the determinant of **M** is equal to 0; i.e.,

$$|\mathbf{M}| = \begin{vmatrix} -k_{1\to2} - E & k_{2\to1} \\ k_{1\to2} & -k_{2\to1} - E \end{vmatrix}$$
$$= \begin{vmatrix} -4.6 \times 10^{-7} - E & 0.25 \\ 4.6 \times 10^{-7} & -0.25 - E \end{vmatrix} = 0$$

(4.4.9)

Recall that the determinant of a 2×2 matrix is obtained by simply cross-multiplying and taking the difference between the two products. Doing this, we obtain a quadratic equation in E:

$$E^2 + (0.25)(4.6 \times 10^{-7}) + E(0.25 + 4.6 \times 10^{-7})$$
$$- (0.25)(4.6 \times 10^{-7}) = 0 \qquad (4.4.10)$$

which has two solutions:

$$E_1 = 0$$
$$E_2 = -k_{1\to2} - k_{2\to1} \qquad (4.4.11)$$
$$= (-4.6 \times 10^{-7} - 0.25) \text{ years}^{-1}$$

We now turn to the task of solving for Φ_1 and Φ_2. Since we have two solutions for E, we will have to solve for the pair of Φ's twice.

4.4.1. Solution 1

For our first solution, $E = E_1 = 0$. Substituting this E-value into Equations (4.4.4) yields

$$\frac{\Phi_2{}^1}{\Phi_1{}^1} = \frac{k_{1\rightarrow 2}}{k_{2\rightarrow 1}} \tag{4.4.12}$$

where the superscript "1" on Φ is used to indicate that these constants are associated with the first solution, E_1. Now substituting this result into Equations (4.4.3), we obtain the first solution to our system of differential equations:

$$C_1(t) = \Phi_1{}^1$$
$$\tag{4.4.13}$$
$$C_2(t) = \Phi_1{}^1 \frac{k_{1\rightarrow 2}}{k_{2\rightarrow 1}}$$

Note that since $E = E_1 = 0$ for this solution, both C_1 and C_2 are constants and do not vary with time.

4.4.2. Solution 2

The second solution has $E = E_2 = (-k_{1\rightarrow 2} - k_{2\rightarrow 1})$. Substituting this E-value into Equations (4.4.4), yields the condition that

$$\Phi_2{}^2 = -\Phi_1{}^2 \tag{4.4.14}$$

Thus our second solution is given by

$$C_1(t) = \Phi_1{}^2 e^{-(k_{1\rightarrow 2} + k_{2\rightarrow 1})t}$$
$$\tag{4.4.15}$$
$$C_2(t) = -\Phi_1{}^2 e^{-(k_{1\rightarrow 2} + k_{2\rightarrow 1})t}$$

4.4.3. The Complete General Solution

The complete general solution to our problem is now obtained through a linear combination of these two solutions. Thus,

$$C_1(t) = (A_1\Phi_1{}^1) + (A_2\Phi_1{}^2)e^{-(k_{1\rightarrow 2} + k_{2\rightarrow 1})t}$$
$$\tag{4.4.16}$$
$$C_2(t) = \frac{k_{1\rightarrow 2}}{k_{2\rightarrow 1}}(A_1\Phi_1{}^1) - (A_2\Phi_1{}^2)e^{-(k_{1\rightarrow 2} + k_{2\rightarrow 1})t}$$

where A_1 and A_2 are weighting factors for combining the two specific solutions into a single general solution. The products $A_1\Phi_1{}^1$ and $A_2\Phi_1{}^2$ in Equation (4.4.16) are constants that must be determined by the initial conditions specified for the system. Since, we lose no generality by representing the product of each of the two constants as a single constant, we can simplify Equation (4.4.16) by writing

$$C_1(t) = Q_1 + Q_2 e^{-(k_{1\rightarrow 2} + k_{2\rightarrow 1})t}$$
$$\tag{4.4.17}$$
$$C_2(t) = \frac{k_{1\rightarrow 2}}{k_{2\rightarrow 1}}Q_1 - Q_2 e^{-(k_{1\rightarrow 2} + k_{2\rightarrow 1})t}$$

where Q_1 and Q_2 are constants equal to the products of $A_1\Phi_1{}^1$ and $A_2\Phi_1{}^2$, respectively. In the examples below, we illustrate how these equations can be used to analyze the behavior of the cycle.

4.4.4. Example 1—The Steady-State Solution

In our first example we assume that the initial populations of the two reservoirs are equal to the steady-state populations we derived in Section 4.3; that is,

$$C_1(0) = 5.5 \times 10^9 \text{ people}$$
$$C_2(0) = 1 \times 10^4 \text{ people}$$

(4.4.18)

The constants Q_1 and Q_2 can now be derived by demanding that the general solution in Equation (4.4.17) gives the initial populations in Equation (4.4.18) when t is set equal to 0. Thus we demand that

$$C_1(0) = Q_1 + Q_2 = 5.5 \times 10^9 \text{ people}$$
$$C_2(0) = \frac{4.6 \times 10^{-7}}{0.25} Q_1 - Q_2 = 10^4 \text{ people}$$

(4.4.19)

Solving, we find that $Q_1 = 5.5 \times 10^9$ people and $Q_2 = 0$. And substituting these values back into Equation (4.4.17), we obtain the particular solution appropriate for the initial conditions specified above:

$$C_1(t) = 5.5 \times 10^9 \text{ people}$$
$$C_2(t) = 1 \times 10^4 \text{ people}$$

(4.4.20)

These, of course, are just the populations we initially specified. Thus, we see that if we start with populations equal to the steady-state populations, we get back the trivial solution that the populations remain fixed, for all time, at these initial steady-state populations.

4.4.5. Example 2—Initial State at the Inception of the University

In our second example we begin at the inception of the university, when all the people reside in the world and no one has yet entered the university. Thus,

$$C_1(0) = 5.50001 \times 10^9$$
$$C_2(0) = 0$$

(4.4.21)

Following the same method we adopted in Example 1, we obtain particular solutions that, in this case, are time-dependent:

$$C_1(t) = (5.50001 \times 10^9 - 1 \times 10^4) + 1 \times 10^4 \, e^{-(0.25 + 4.6 \times 10-7)t}$$
$$\approx 5.5 \times 10^9 \text{ people}$$
$$C_2(t) = 1 \times 10^4 \, (1 - e^{-(0.25 + 4.6 \times 10-7)t})$$
$$\approx 1 \times 10^4 \, (1 - e^{-0.25t}) \text{ people}$$

(4.4.22)

We find that the world population is hardly affected by the growth of the university. On the other hand, the university population grows to its steady-state population at a rate characterized by the sum of $k_{1\to2}$ and $k_{2\to1}$. In this specific case, $(k_{1\to2} + k_{2\to1}) \sim 0.25$ years^{-1}. The resulting variations in C_1 and C_2 are illustrated in Figure 4.2.

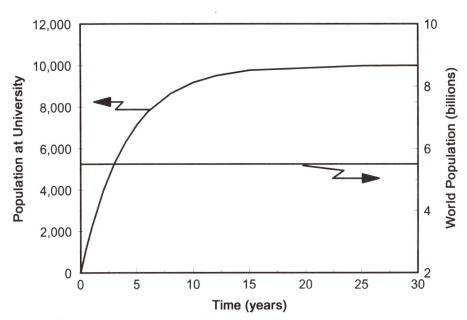

Figure 4.2. The solution to the university–world cycle for Example 2, which has as its initial condition the inception of the university. Note the negligible impact of the university in the world population and the rapid approach to steady state.

4.4.6. Example 3—Perturbation Experiment

In the third example, we examine how the system responds to a perturbation. Suppose that the system has equilibrated to the steady-state populations derived in Section 4.3. Suddenly the university's course requirements for graduation are radically altered. Because of this change, $\tau_{2\rightarrow1}$, the residence time for students at the university increases from 4 years to 1 million years. Thus we now have that

$$k_{2\rightarrow1} = \frac{1}{10^6 \text{ years}} = 10^{-6} \text{ years}^{-1} \tag{4.4.23}$$

with $k_{1\rightarrow2}$ unchanged from its original value of 4.6×10^{-7} years^{-1}. Because $k_{2\rightarrow1}$ has changed without a corresponding change $k_{1\rightarrow2}$, the populations derived in Section 4.3 that had been in steady state are no longer in steady state.

To determine how the system will respond to this change in $k_{2\rightarrow1}$, we must first rederive the general solution for $C_1(t)$ and $C_2(t)$ with this value for $k_{2\rightarrow1}$. Following the same procedure we outlined in the previous sections , we find that

$$C_1(t) = Q_1 + Q_2 e^{-(1.46 \times 10^{-6} t)}$$
$$C_2(t) = 0.46 \, Q_1 - Q_2 e^{-(1.46 \times 10^{-6} t)} \tag{4.4.24}$$

We must now solve for Q_1 and Q_2, by applying the appropriate initial conditions. Since we had specified that the populations were at their original steady-state values when the change in $k_{2\rightarrow1}$ was made, the appropriate initial conditions are that $C_1(0) = 5.5 \times 10^9$ people and $C_2(0)1 \times 10^4$ people. Applying these conditions, we find that

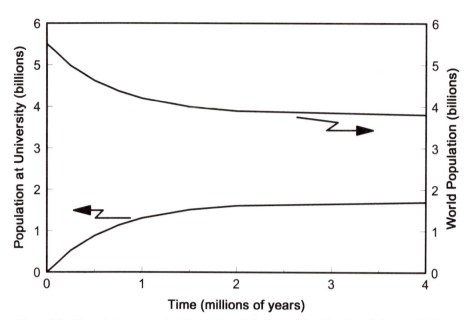

Figure 4.3. The solution to the university–world cycle for Example 3, with a dramatic increase in University residence time at $t = 0$. Note the substantial increase in the university population and the much longer time required in this case to reach steady state.

$$C_1(t) = [3.8 \times 10^9 + 1.7 \times 10^9 \, e^{-(1.46 \times 10^{-6} \, t)}]$$
$$C_2(t) = [1.7 \times 10^9 \, (1 - e^{-(1.46 \times 10^{-6} \, t)})]$$

(4.4.25)

As illustrated in Figure 4.3, we find in this case that the world population slowly decays from its initial value of 5.5 billion to a new steady state of 3.8 billion, while the university population grows to its new steady-state value of 1.7 billion. (Apparently the new course regulations are going to require that the University administration build a lot of new dormitories to house all the extra students that will be residing there. The good news for the university administration is that they will have a long time to build these new dormitories. Although the old cycle with a 4-year university residence time required only about 4 years to approach steady state, the new cycle takes about 700,000 years. Is there any way for you to have predicted this much longer equilibration time without first solving the differential equations?)

4.4.7. Summary

This simple cycle and the examples given above illustrate a number of characteristics common to the biogeochemical cycles we will be studying here:

• C_i is used to denote the amount of an element in the ith reservoir. While in our simple, two-box, university–world cycle these C_i's represented populations, they will generally represent the number of moles or grams of an element in a given reservoir in the cases of the actual biogeochemical cycles we will study.

• The parameter $k_{i \to j}$ is used to denote the transfer coefficient from the ith to the jth reservoir. These coefficients will always have units of inverse time, usually years^{-1}. In general, we can think of the k's as being related to the reciprocals of the residence times in the various reservoirs. For the simple, two-box, university–world cycle analyzed in this section, the relationship between the residence times for the two reservoirs and the two k values is straightforward. For more complicated systems, involving three or more reservoirs, τ_i, the residence time in the ith reservoir, is given by

$$\tau_i = \left(\sum_{\substack{j \neq i}}^{N} k_{i \to j} \right)^{-1} \tag{4.4.18}$$

where N is the number of reservoirs in the system.

• In our simple, two-box, university–world cycle, we needed two differential equations. More generally, a system containing N reservoirs will require N-coupled first-order, homogeneous differential equations. Just as our two-box system yielded two independent solutions, a system of N boxes with N differential equations will yield N different, independent solutions of the form Ae^{Et}, where A is a constant. Of these N solutions, one will be a time-independent solution, with $E = 0$. The other $N - 1$ solutions will each be characterized by a different E value. All of these $(N - 1)$ E's will be less than zero (thus guaranteeing that the solution will not diverge as $t \to \infty$).

• The response time of the system is defined as the characteristic e-folding time for the system to approach its steady-state solution. This time is given by the inverse of the E with the smallest, nonzero magnitude. Thus in our simple, two-box, university–world cycle, the response time is given by

$$\tau_{\text{response}} = \frac{1}{k_{1 \to 2} + k_{2 \to 1}} \tag{4.4.19}$$

which turns out to be 4 years in our initial examples with a 4-year university residence time and 0.68 million years in the third example with a 1-million-year university residence time.

• The complete general solution for each reservoir of an N-box system is a linear combination of the N different solutions obtained from solving the N-coupled, differential equations. This combination will consist of $N - 1$ terms of the form $Q_i \exp(E_i t)$ and one time-independent term arising from the $E = 0$ solution. The exact combination of these individual solutions is determined by the initial conditions specified for the system. Since the $N - 1$ E_i's are all less than zero, these terms will all decay to zero as $t \to \infty$, leaving only the time-independent term. Thus the $\exp(E_i t)$ terms in the solution describe the transient response of the system, and the time-independent term represents the steady-state solution.

4.5. THE EIGENVECTOR / EIGENVALUE SOLUTION METHOD FOR BIOGEOCHEMICAL CYCLES

The method we used to solve the two-box university–world cycle in Section 4.4 has a specific mathematical name. It's called the eigenvector/eigenvector method of solving coupled differential equations. While it was not necessary for us to use this specific method with its vector-matrix formulation to solve our simple 2-box university–world cycle, the method turns out to be extremely powerful when we are dealing with more complex cycles involving three or more reservoirs. To illustrate this, we now turn to the more general problem of simulating a biogeochemical cycle having N reservoirs, where N can be any positive number greater than 1.

4.5.1. Statement of the General N-Box Problem

Suppose the following:

1. We have an N-box system.
2. Each of the boxes or reservoirs contains $C_i(t)$ grams or moles of material at time, t, where $i = 1, 2, \ldots, N$.
3. At time $t = 0$, $C_i(0) = C_{i0}$, for $i = 1, 2, \ldots, N$.
4. The rate of transfer of material from the ith reservoir to the jth reservoir is given by $F_{i \to j} = k_{i \to j} C_i$.

The problem: Solve for $C_i(t)$, for $i = 1, 2, \ldots, N$.

4.5.2. Setting Up the Problem in Vector/Matrix Form

As in discussed in Section 4.4, our N-box system will have N first-order differential equations. These equations will take the following form:

$$\frac{dC_1(t)}{dt} = k_{2 \to 1} C_2(t) + k_{3 \to 1} C_3(t) + \cdots + k_{N \to 1} C_N(t)$$

$$-k_{1 \to 2} C_1(t) - k_{1 \to 3} C_1(t) - \cdots - k_{1 \to N} C_1(t)$$

$$\frac{dC_2(t)}{dt} = k_{1 \to 2} C_1(t) + k_{3 \to 2} C_3(t) + \cdots + k_{N \to 2} C_N(t)$$

$$-k_{2 \to 1} C_2(t) - k_{2 \to 3} C_2(t) - \cdots - k_{2 \to N} C_2(t)$$

$$\frac{dC_3(t)}{dt} = k_{1 \to 3} C_1(t) + k_{2 \to 3} C_2(t) + \cdots + k_{N \to 3} C_N(t) \qquad (4.5.1)$$

$$-k_{3 \to 1} C_3(t) - k_{3 \to 2} C_3(t) - \cdots - k_{3 \to N} C_3(t)$$

$$\vdots \qquad \vdots \qquad \vdots \qquad \vdots$$

$$\frac{dC_N(t)}{dt} = k_{1 \to N} C_1(t) + k_{2 \to N} C_2(t) + \cdots + k_{(N-1) \to N} C_{N-1}(t)$$

$$-k_{N \to 1} C_N(t) - k_{N \to 2} C_N(t) - \cdots - k_{N \to (N-1)} C_N(t)$$

Note that each of the equations in (4.5.1) is composed of two sets of terms: (i) a series of source terms on the first line of the equation which represent the flow of material into the box and (ii) a series of sink or loss terms on the second line of the equation which represent the flow of material out of the box.

Equations (4.5.1) can be simplified somewhat by collecting all the sink terms in each equation into a single summation over a dummy index j and then rearranging terms:

$$\frac{dC_1(t)}{dt} = -\sum_{j \neq 1}^{N} [k_{1 \to j}C_1(t)] + k_{2 \to 1}C_2(t) + k_{3 \to 1}C_3(t) + \cdots + k_{N \to 1}C_N(t)$$

$$\frac{dC_2(t)}{dt} = k_{1 \to 2}C_1(t) - \sum_{j \neq 2}^{N} [k_{2 \to j}C_2(t)] + k_{3 \to 2}C_3(t) + \cdots + k_{N \to 2}C_N(t)$$

$$\frac{dC_3(t)}{dt} = k_{1 \to 3}C_1(t) + k_{2 \to 3}C_2(t) - \sum_{j \neq 3}^{N} [k_{3 \to j}C_3(t)] + \cdots + k_{N \to 3}C_N(t)$$

$$\cdot \qquad \cdot \qquad \cdot \qquad \cdot$$

$$\cdot \qquad \cdot \qquad \cdot \qquad \cdot$$

$$\cdot \qquad \cdot \qquad \cdot \qquad \cdot$$

$$\frac{dC_N(t)}{dt} = k_{1 \to N}C_1(t) + k_{2 \to N}C_2(t) + \cdots + k_{(N-1) \to N}C_{N-1}(t) - \sum_{j=1}^{N-1} [k_{N \to j}C_N(t)]$$

$$(4.5.2)$$

Inspection of (4.5.2) will reveal that for each of the N equations, the time rate of change in a given reservoir amount, C_i (for $i = 1$ to N) is equated to a linear combination of the N reservoir amounts. In other words, dC_i/dt is equal to a summation of all C_j's for $j = 1$ to N. Further note that in this summation, positive coefficients multiply the C_j's when $j \neq i$ and negative coefficients when $j = i$. Because of the common form of each of the equations in (4.5.2), they can be expressed in more compact form by recasting the N equations into a single equation involving the index i, which varies from 1 to N, and a matrix $\mathbf{K}$:

$$\frac{dC_i(t)}{dt} = \sum_{j=i}^{N} K_{ij}C_j(t), \qquad i = 1, 2, \ldots, N \qquad (4.5.3a)$$

where

$$K_{ij} = k_{j \to i} \qquad \text{when } i \neq j \qquad (4.3.2b)$$

and

$$K_{ii} = -\left(\sum_{j \neq i}^{N} k_{i \to j} \right) \qquad (4.5.3c)$$

The reader should make note of the change in notation used for the uppercase K as opposed to that used for the lowercase k. We drop the arrow from the subscript in K and we reverse the order of the i and j indices. As will become apparent below, we do this to facilitate our adapting K to matrix notation.

The final step in our development of the differential equations for a N-box system is to note that Equations (4.5.3) can be cast in vector–matrix form:

$$\frac{d\vec{C}(t)}{dt} = \mathbf{K}\vec{C}(t) \tag{4.5.4a}$$

where

$$\vec{C}(t) \equiv \text{an } N\text{-dimensional vector} = \begin{matrix} C_1 \\ C_2 \\ C_3 \\ \cdots \\ C_{N \to 1} \\ C_N \end{matrix} \tag{4.5.4b}$$

and

$$\mathbf{K} \equiv \text{an } N \times N \text{ matrix} = \begin{matrix} K_{11} & K_{12} & \cdots & K_{1N} \\ K_{21} & K_{22} & \cdots & K_{2N} \\ K_{31} & K_{32} & \cdots & K_{3N} \\ \cdots & \cdots & \cdots & \cdots \\ K_{N1} & K_{N2} & \cdots & K_{NN} \end{matrix} \tag{4.5.4c}$$

and the K_{ji}'s are related to the transfer coefficients, $k_{i \to j}$, by Equations (4.5.3b) and (4.5.3c).

The solution to Equation (4.5.3) or (4.5.4) is obtained in essentially the same manner that we solved the two-box problem in Section 4.4, only now we work with vectors and matrices instead of scalars. As in our two-box problem, we assume a solution with an exponential dependence on time:

$$\vec{C}(t) = \vec{\Phi}\exp(Et) \tag{4.5.5a}$$

where E is a constant

$$\vec{\Phi} \equiv \text{an } N\text{-dimensional vector} = \begin{matrix} \Phi_1 \\ \Phi_2 \\ \cdots \\ \Phi_{N-1} \\ \Phi_N \end{matrix} \tag{4.5.5b}$$

and the Φ_i are constants (and are equivalent to the Φ_1 and Φ_2 used in the two-box problem). As illustrated below, the solution to this system consists of N independent solutions, each having a different E value and a different vector $\vec{\Phi}$. By convention the set of N different E values are referred to as the *eigenvalues* of the $\mathbf{K}$ matrix, and the set of N $\vec{\Phi}$ vectors are referred to as the *eigenvectors* of the $\mathbf{K}$ matrix.

4.5.3. Obtaining the General Solution to the Eigenvector/ Eigenvalue Problem

The general solution to the matrix–differential equation stated in Equation (4.5.4) can be obtained by substituting Equation (4.5.5) into (4.5.4). This yields N equations of the form

$$\frac{dC_i(t)}{dt} = E\Phi_i \exp(Et) = \sum_{j=i}^{N} K_{ij}\Phi_j \exp(Et), \qquad i = 1, 2, \ldots, N \quad (4.5.6)$$

As we did in the 2-box model in Section 4.4, the exponential term in these equations can be factored out and the right-hand-side terms can be brought to the left-hand-side. We are left with N algebraic equations of the form:

$$\sum_{j=1}^{N} (E\delta_{ij} - K_{ij})\Phi_j = 0, \qquad i = 1, 2, \ldots, N \quad (4.5.7)$$

where δ_{ij} is the so-called *delta function* (i.e., $\delta_{ij} = 1$, when $i = j$, and $\delta_{ij} = 0$, when $i \neq j$). Multiplying this result by a minus sign and recasting the equation in vector-matrix form, we obtain an equation which looks identical to Equation (4.4.6):

$$(\mathbf{M})\vec{\Phi} = 0 \quad (4.5.8)$$

In this case, however, the $\mathbf{M}$ matrix is defined by

$$\mathbf{M} = \begin{matrix} K_{11} - E & K_{12} & \cdots & K_{1N} \\ K_{21} & K_{22} - E & \cdots & K_{2N} \\ \cdots & \cdots & \cdots & \cdots \\ K_{N1} & K_{N2} & \cdots & K_{NN} - E \end{matrix} \quad (4.5.9)$$

While similar to the definition for the $\mathbf{M}$ matrix that we used in our 2-box University–World cycle (Equation 4.4.7) there are some subtle differences. In Equation (4.4.7), lower-case k-values were used, while here we use the upper-case K-values. Using upper-case K's, gives us standard matrix notation for the subscripts of each of the elements in the $\mathbf{M}$ matrix. It also results in the appearance of "$+$" signs on the diagonal elements of the matrix where "$-$" signs had appeared previously.

A non-trivial solution to the matrix equation in (4.5.8) is one where the determinant of the $\mathbf{M}$ matrix is equal to 0. Calculation of the determinant of the $\mathbf{M}$ matrix and setting it equal to zero, yields a polynomial equation in E of order N. This polynomial equation has N solutions ($E_1, E_2, \ldots, E_N$) which, as noted earlier, are refered to as the eigenvalues of the $\mathbf{K}$ matrix.

Associated with each of the N eigenvalues there is an eigenvector, $\vec{\Phi}$. The eigenvector associated with the nth eigenvalue, E_n, is

$$\vec{\Phi}_n = \begin{matrix} \Phi_{1n} \\ \Phi_{2n} \\ \cdots \\ \Phi_{Nn} \end{matrix} \quad (4.5.10)$$

These eigenvectors are, in turn, obtained by substitution into Equation (4.5.7)

$$\sum_{j=1}^{N} (E_n\delta_{ij} - K_{ij})\Phi_{jn} = 0, \qquad i = 1, 2, \ldots, N \quad (4.5.11)$$

and solving this set of N-coupled linear equations for Φ_{jn}. A final simplification to the formulation of the solution is typically made by defining an "eigenvector" matrix

$$
\Psi = \begin{matrix}
\Phi_{11} & \Phi_{12} & \cdots & \Phi_{1N} \\
\Phi_{21} & \Phi_{22} & \cdots & \Phi_{2N} \\
\cdots & \cdots & & \\
\Phi_{N1} & \Phi_{N2} & \cdots & \Phi_{NN}
\end{matrix} \tag{4.5.12}
$$

where each of the columns in this matrix is one of the N eigenvectors defined in Equation (4.5.6).

We thus have N solutions to our problem consisting of N eigenvalues and an $N \times N$ eigenvector matrix. It should be noted that both the eigenvalues and the eigenvector matrix are uniquely and completely defined by the $\mathbf{K}$ matrix.

As in our two-box problem, the complete, general solution to our N-box problem is obtained through a linear combination of all N solutions; that is,

$$
C_i(t) = \sum_{n=1}^{N} a_n \Psi_{in} \exp(E_n t) \tag{4.5.13}
$$

where the a_n's are linear weighting factors for each of the eigenvalues. This solution is further simplified by rewriting it as

$$
C_i(t) = \sum_{n=1}^{N} \mathbf{Q}_{in} \exp(E_n t) \tag{4.5.14}
$$

where $\mathbf{Q}_{in} = a_n \Psi_{in}$ combines the weighting factor and the eigenvector components into a single constant. In the section below, we outline how the $\mathbf{Q}$'s are obtained from the initial conditions.

4.5.4. Obtaining the Particular Solution by Applying Initial Conditions

The last step in obtaining the solution to our eigenvector/eigenvalue problem is to solve for the Q's in Equation (4.5.14). Just as in the solution for the two-box model presented in Section 4.4, this last step is accomplished by demanding that Equation (4.5.14) yield the appropriate initial values (i.e., C_{i0}) at time $t = 0$; that is,

$$
C_i(0) = C_{i0} = \sum_{n=1}^{N} \mathbf{Q}_{in} = \sum_{n=1}^{N} \Psi_{in} a_n, \qquad i = 1, 2, \ldots, N \tag{4.5.15}
$$

Equation (4.5.15) yields N linear, coupled equations whose solution yields the appropriate values for a_n and, by extension, the $\mathbf{Q}$-matrix. In practice, for $N \geq 2$, this solution is most readily obtained by rewriting the equation in matrix form and carrying out a matrix inversion; that is,

$$
\vec{C}_{i0} = \Psi \vec{a} \tag{4.5.16}
$$

whose solution is

$$
\vec{a} = \Psi^{-1} \vec{C}_{i0} \tag{4.5.17}
$$

4.5.5. Summary of Eigenvalue/Eigenvector Method

The simulation of a biogeochemical cycle using the eigenvector/eigenvalue method involves six basic steps:

1. Specify the N reservoirs and the appropriate initial conditions, C_{i0}, and transfer coefficients, $k_{i \to j}$.

2. Set up the **K** matrix from the $k_{i \to j}$'s using Equations (4.5.3b) and (4.5.3c).

3. Use any one of several mathematical or numerical methods, and solve for the eigenvalues $(E_1, E_2, \ldots, E_N)$ and eigenvector matrix (Ψ) of the **K** matrix.

4. From the initial conditions, determine the weighting factors $(a_1, a_2, \ldots, a_N)$ from Equation (4.5.16) through matrix inversion.

5. Determine Q matrix from the eigenvector matrix and weighting factors, where $Q_{in} = a_n \Psi_{in}$; and

6. Write out the final solution as

$$C_i(t) = Q_{i1} \exp(E_1 t) + Q_{i2} \exp(E_2 t) + \cdots + Q_{iN} \exp(E_N t) \qquad (4.5.18)$$

As noted earlier, the solution involves N eigenvalues, E_n. Of these N eigenvalues, $N - 1$ of them will be negative, and the remaining one will be equal to zero. Because the eigenvalues appear in the solution as exponential factors, all the terms of the solution with negative eigenvalues will decay with time, leaving the term with the zero eigenvalue as the steady-state solution.

4.6. RECIPE FOR MODELING BIOGEOCHEMICAL CYCLES USING *BOXES*

The mathematical tasks outlined in Section 4.5 for using the eigenvector/eigenvalue method to integrate the differential equations that describe biogeochemical cycles are, in principle, straightforward. In practice, however, they can be arduous, especially if one of us dealing with a cycle involving three or more boxes.

Because of computers, however, much of the tedium associated with using the eigenvector/eigenvalue method is no longer necessary. Instead of having to solve the numerous algebraic equations associated with the method by hand, we can now simply program a computer to do it for us. To make the task even easier, a computer program entitled *BOXES* has been included with the textbook. With appropriate input, *BOXES* will help you set up a model of any closed biogeochemical cycle (provided $N \le 10$), solve for the reservoir amounts as a function of time, and aid in the analysis of the results by displaying the solution in both tabular and graphical form. To use *BOXES* to simulate a biogeochemical cycle, you must first carry out three preliminary steps:

Preliminary Step 1. Set up a basic schematic diagram of your cycle (as in Figure 4.1 for the university–world cycle). The diagram should identify the N reservoirs or boxes of the cycle and the various pathways by which material is exchanged between the reservoirs.

Preliminary Step 2. Calculate the transfer coefficients, $k_{i \to j}$, for each pathway. Recall that the rate of transfer from the ith to the jth reservoir is given by $[k_{i \to j} C_i(t)]$. For an N-reservoir system, there will be $N(N - 1)$ k's, although most likely some of them will be equal to 0.

Preliminary Step 3. Determine the initial reservoir amounts; that is, $C_i(t = 0)$.

Once these steps are completed, you are ready to run *BOXES*. The code is menu-driven with context-sensitive help available at any time by pressing the "F1" key or using the mouse to activate the "Help" command on the menu bar that appears along the top of the screen. After starting *BOXES*, you must first input the information you gathered in the preliminary steps. This is accomplished by invoking the "Boxes" command and then clicking sequentially on each Box # that appears in the dialog box that appears.. When you click on a Box #, a new dialog box—the "Box Properties Dialog Box"—will appear with blank spaces for the "BOX name," initial "contents," and k values (see Text Box 4.1). Using the mouse or "Enter" to move from space to space within this dialog box, you simply enter the appropriate name, initial contents, and k

---

TEXT BOX 4.1	THE RELATIONSHIP BETWEEN THE $K_{i \to j}$'S AND THE K MATRIX, AND THEIR USE IN *BOXES*

The task of setting up and then solving the differential equations that describe a biogeochemical cycle will be a good deal easier if you remember the relationship between the $k_{i \to j}$'s and the **K** matrix. Recall that each $k_{i \to j}$ represents the exchange coefficient for the flow of material **from** reservoir "i" and **to** reservoir "j". The **K** matrix, on the other hand, is nothing more than a collection of all the $k_{i \to j}$ values. However, the ordering of this collection is very specific. For an N-box cycle, this collection will take the form of an $N \times N$ matrix with each vertical column of the matrix containing a sequencing of the $k_{i \to j}$ values for a single value of "i" and all possible values of "j". The diagonal element in this column (i.e., the K_{ii} term) is the negative sum of the $k_{i \to j}$ values contained in the column. In other words, the mth column of an N-box cycle will have the following form:

$$k_{m \to 1}$$
$$k_{m \to 2}$$
$$k_{m \to 3}$$
$$\cdots$$
$$\cdots$$
$$\cdots$$
$$-\Sigma(k_{m \to j})$$
$$\cdots$$
$$\cdots$$
$$\cdots$$
$$k_{m \to N-1}$$
$$k_{m \to N}$$

To make things as easy as possible, the **K** matrix is entered into *BOXES* as columns of $k_{i \to j}$ values just as they appear above. For each box or reservoir, a "Box Properties Dialog Box" provides blank spaces for entering the coefficient for the transfer of material from that box to all the other boxes. The field for entering the diagonal element cannot be entered, but is instead automatically calculated by *BOXES* as the negative sum of the other $k_{i \to j}$ values that are entered into that dialog box.

values and then click on "OK" to exit the dialog box and return to the main screen. After you have finished entering the data for all the boxes or reservoirs, you can then use the "Run" command to obtain a solution, the "Output" command to view the results in text format, the "Plot" command to view the results graphically, and the "File" command to save and load data files. Some simple problems that will give you an opportunity to familiarize yourself with the operation of *BOXES* are given at the end of the chapter.

4.7. CONCLUSION

The linear box model approach for analyzing global biogeochemical cycles will prove to be extremely useful in the chapters that follow. Nevertheless, it is important to bear in mind that this approach simplifies a very complex system and is, therefore, not without its limitations. Probably the most serious of these limitations is the assumption implicit in the approach that all transfer rates depend linearly on the abundance of a single element in a single reservoir. In fact, most biogeochemical processes are considerably more complex than that. Thus, as we study the global biogeochemical cycles of the elements in the ensuing chapters and use *BOXES* to simulate these cycles, it is important that we always examine the results critically and objectively to make sure that the results are reasonable.

We have now completed the introductory portion of our subject. We have reviewed the basic principles of chemical thermodynamics and the spheres of the earth system, and we have developed a mathematical formulation (and introduced computer software based on this formulation) that is capable of simulating many of the essential features of biogeochemical cycles. Now we are ready to investigate the biogeochemical cycles themselves.

SUGGESTED READING

Hildebrand, F. B., *Introduction to Numerical Analysis*, 2nd Edition, McGraw-Hill, New York, 1974.
Lasaga, A. C., Dynamic treatment of geochemical cycles: Global kinetics, in *Kinetics of Geochemical Processes, Reviews of Mineralogy*, **8**, 69–109, 1988.
Watkins, D. S., *Fundamentals of Matrix Computations*, John Wiley, New York, 1991.

PROBLEMS

1. Because of a global shortage of biogeochemists, it is proposed that the University of Biogeochemistry we considered in Section 4.3 double its acceptance rate of new students. In order to plan for the building of new dormitories, the Board of Directors want to know how many students will eventually be at the university and how long it will take for the student population to reach this new number, should they decide to adopt this proposal. The Board comes to you, as an expert in biogeochemistry, for the answers to these questions. Use your intuition to guess at the answers and then use *BOXES* to confirm its accuracy.

2. Instead of doubling its acceptance rate as proposed in Problem 1, the Board of Directors of the University of Biogeochemistry decide to add a 6-year Ph.D. program to their institution. Based on an assessment of the abilities of their undergraduates, they decide that 25% of the students that graduate from their university as undergraduates should be accepted to their Ph.D. program. Use *BOXES* to predict the temporal evolution of the Ph.D. student population at the university.

The Global Phosphorus Cycle

5

> "... Phosphorus has a very beautiful name (it means 'bringer of light'), it is phosphorescent, it's in the brain; ... without phosphorus plants do not grow; ... it is in the tips of matches, and girls driven desperate by love ate them to commit suicide; it is in will-o'-the-wisps, putrid flames fleeing before the wayfarer."
>
> P. Levi, *The Periodic Table*, Schocken Books, New York, 1984.

5.1. INTRODUCTION

Phosphorus (P) is a nonmetallic element in Group VA of the Periodic Table (Figure 5.1) with an atomic number of 15 and an atomic weight of 30.975. The element was discovered in 1669 by H. Brand. In 1771, K. W. Schelle found traces of it in bone ash, thus establishing phosphorus' presence in living organisms. As it turns out, P is not merely present in living organisms, it is a critical component. Although it amounts to only about 1% by weight of most organic matter, it plays a central role in the storage, transport, and utilization of energy within cells (see Text Box 5.1). Moreover, because of the element's propensity to form insoluble compounds in natural environments, P frequently limits the rate of photosynthetic productivity in many ecosystems, especially aquatic ones, despite the fact that it is the tenth most abundant element on the earth. For these reasons, the biogeochemical cycle of P is fundamental to the survival of the biosphere. We begin our examination of this cycle by briefly reviewing some the basic features of the thermochemistry of P.

5.2. P REDOX CHEMISTRY

Before we can construct a cycle of P, we need to have some idea of the kinds of chemical forms that the element might take in the earth system. This requires that we consider P's redox chemistry. Now, if we go back to Chapter 2 and review Table 2.2, we see that the element P is found in four oxidation states: $-3, 0, +3,$ and $+5$. As sum-

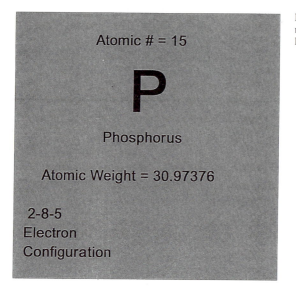

Atomic # = 15

P

Phosphorus

Atomic Weight = 30.97376

2-8-5
Electron
Configuration

Figure 5.1. Phosphorus is a non-metallic element in Group VA of the Periodic Table.

marized in Figure 5.3, the chemical compounds associated with these oxidation states include: *phosphine* (PH_3) for the -3 oxidation state, elemental phosphorus (often found in its amorphous state, P_4), phosphorus acid and its conjugate bases (H_3PO_3, $H_2PO_3^-$, HPO_3^{2-}) for the $+3$ oxidation state, and phosphoric acid and its conjugate

TEXT BOX 5.1	**THE ROLE OF P IN CELLULAR METABOLISM**

Phosphorus' key role in the biosphere arises from its presence in two compounds found within cells: *adenosine di-* and *triphosphate* (ADP and ATP). These two compounds are, in turn, composed of two organic compounds, adenine and ribose, and two phosphates in the case of ADP and three phosphates in the case of ATP. When ATP is hydrolyzed (i.e., when it reacts with water), it yields ADP, phosphate, and energy; conversely, with the addition of energy, ADP and phosphate can be combined to produce ATP; that is,

$$ATP + H_2O \rightleftharpoons ADP + PO_4^{2-} + Energy$$

Through this reaction couplet, cells are able to store energy, transport it in the form of ATP, and then release this energy at the time and location that it is needed. For instance, within *mitochondria*, which are the respiratory centers of a cell, ATP serves as a chemical intermediary in the conversion of carbohydrates and fats into energy that can then be transported to and used by other parts of the cell (see Figure 5.2). Within the *chloroplast* of green plants, where photosynthesis is accomplished, the energy from sunlight is used to fuel a series of enzymatic reactions that also produce ATP from ADP. The energy stored in this ATP is eventually used to complete the conversion of the initial products of photosynthesis into carbohydrates.

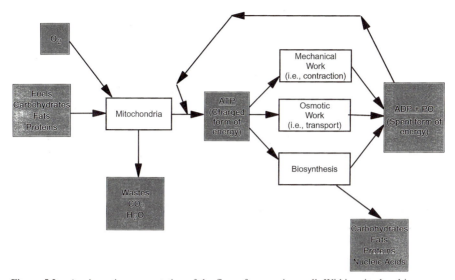

Figure 5.2. A schematic representation of the flow of energy in a cell. Within mitochondria, energy released from reactions between O_2 and carbohydrates, fats, and proteins is used to produce ATP from ADP and phosphate (PO_4^{2-}). The ATP is then transported to other parts of the cell where its chemical energy is released to support a variety of metabolic processes, including mechanical work, osmotic work, and biosynthesis. The breakdown of the ATP produces ADP and PO_4^{2-}, which pass back to the mitochondria where they are regenerated into ATP. (After Swanson, *The Cell*, Prentice-Hall, Englewood Cliffs, NJ, 1969, 150 pages.)

bases, sometimes referred to as the *orthophosphates* (H_3PO_4, $H_2PO_4^-$, HPO_4^{2-}, PO_4^{3-}), for the +5 oxidation state. There are two important features of these compounds that are worth noting. Firstly, although H_3PO_3 has three protons, it is only a diprotic acid with two, rather than three, conjugate bases. Secondly, of all the compounds listed above, only PH_3 has a significant vapor pressure for atmospheric conditions and is therefore the only compound that has the potential to be a gaseous component of the atmosphere. As we will soon discover, however, PH_3 is not a natural atmospheric constituent because of P's thermodynamic properties.

To determine which of the four oxidation states discussed above might play a role in the biogeochemical cycle of P, we need to examine its pe–pH stability diagram. A rendition of this diagram is presented in Figure 5.4. Note in the figure that only P in the +5 oxidation state (i.e., the orthophosphates) appears within the stability region of water (i.e., between the stability lines of O_2 and H_2). From our discussions in Chapter 2, we know that this means that only P in the +5 oxidation

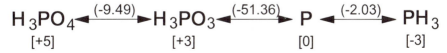

Figure 5.3. The oxidation states of P. The numbers in square brackets beneath each phosphorus bearing compounds indicate the oxidation state of P within the compound, and the numbers in parentheses between the compound indicates the log of the equilibrium constant for the redox half-reaction between the two compounds.

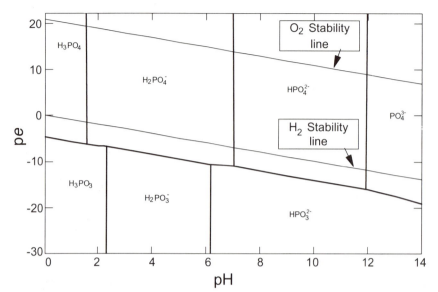

Figure 5.4. The pe–pH stability diagram for P using the procedure outlined in Chapter 2 and the equilibrium constants listed in the Appendix. Note that only P compounds in the +5 oxidation state (i.e., H_3PO_4, $H_2PO_4^-$, HPO_4^{2-}, and PO_4^{3-}) appear within the stability field of H_2O (i.e., below the O_2 stability line and above the H_2 stability line). Hence, only the +5 oxidation state of P need be considered in the biogeochemical cycle of P.

state is stable in an aqueous solution. As it turns out, the +5 oxidation state is essentially the only form of P found in any significant abundance within the earth system (see Table 5.1). In the ocean, dissolved P is always found as some form of PO_4^{2-}. Similarly, within the lithosphere, P occurs in a variety of minerals, but the P in these minerals is virtually always in the form of phosphate. The most common of the P-containing mineral is *apatite*, which has a hexagonal crystal structure and includes *fluorapatite* ($Ca_5(PO_4)_3F$), *hydroxyapatite* ($Ca_5(PO_4)_3OH$), and *chlorapatite* ($Ca_5(PO_4)_3Cl$). Within the biosphere, P is also present—but only as phosphate units within the *nucleic acids* (i.e., DNA and RNA)—as well as ATP and ADP. Because P in the earth system is limited to the +5 oxidation state, the biogeochemical cycle of P does not include a stable gaseous state. P in the atmosphere is found

TABLE 5.1
Biogeochemical Cycle of Phosphorus is Limited to +5 Oxidation State

Earth sphere	Major Phosphorus-Containing Species
Ocean	Orthophosphates{$MgHPO_4$, HPO_4^{2-}, $NaHPO_4^-$, $CaPO_4^-$, $CaHPO_4$} Organic orthophosphate {sugar phosphates, phospholipids}
Lithosphere	Phosphate minerals {Apatite: $Ca_5(PO_4)_3F$, $Ca_5(PO_4)_3OH$, $Ca_5(PO_4)_3Cl$}
Biosphere	Organic phosphates {DNA, RNA, ATP, ADP}
Atmosphere	Phosphate minerals {in wind-blown dust and sand} Orthophosphates {in cloud and rainwater}

either in solid form, as a mineral constituent contained in wind-blown dust and sand, or as dissolved phosphates, in clouds and rainwater (see Table 3.4).

5.3. THE BIOGEOCHEMICAL REACTIONS OF THE P CYCLE

In the previous section we found that the biogeochemical cycle of P is limited to the element in its $+5$ oxidation state (i.e., PO_4^{3-}). Because of the prevalence of phosphate within the hydrosphere, lithosphere, and biosphere, we might expect these spheres to represent significant reservoirs of the element. On the other hand, because there are no stable gaseous P compounds in the $+5$ oxidation state, we might expect the atmosphere to represent a relatively unimportant reservoir for P. The next step in developing our global cycle for P is to identify the processes or biogeochemical reactions that cause P to flow from one sphere of the earth system to another.

5.3.1. Coupling of the P Cycle to the Biosphere—Photosynthesis and Respiration

The flow of P into and out of the biosphere is controlled by the processes of photosynthesis, which incorporates inorganic phosphate into living organisms, and respiration and decay, which reconverts organic phosphate back to its inorganic forms. As discussed in Chapter 2, we will represent these complex processes as two stoichiometric reactions: one for the ocean

$$106CO_2 + 64H_2O + 16NH_3 + H_3PO_4 + h\nu$$
$$\rightleftarrows C_{106}H_{179}O_{68}N_{16}P + 106O_2 \qquad (R3.3)$$

and the other for the continents

$$830CO_2 + 600H_2O + 9NH_3 + H_3PO_4 + h\nu$$
$$\rightleftarrows C_{830}H_{1230}O_{604}N_9P + 830O_2 \qquad (R3.4)$$

In both cases, recall that photosynthesis is represented by the reaction proceeding from left to right whereas respiration and decay are represented by the reaction proceeding from right to left.

The stoichiometric relationship between N and P in the oceanic reaction (R3.3) is interesting to consider in further detail, because it suggests that P plays a profound role in regulating ocean productivity. In the deep ocean, the total concentration of dissolved P is about 2 μM (i.e., 2×10^{-6} moles liter^{-1}), and the concentration of dissolved nitrogen is about 35 μM. Thus the ratio of N:P in the deep ocean is about 17:1. But from (R3.3) we see that the ratio of N:P in marine organisms is quite similar—about 16:1. Marine geochemists do not believe that the similarity between these two ratios is simply a coincidence. Instead, they believe that the N:P ratio of the ocean has been "engineered" by a special class of oceanic biota, called *nitrogen fixers*. These nitrogen fixers, which include *cyanobacteria* or *blue-green algae*, have the ability to convert molecular N_2 (that cannot be used in photosynthesis) into usable nitrogen (i.e., nitrate and ammonium) that can be assimilated in photosynthesis. Because nitrogen fixation requires energy, it is unlikely that these organisms fix nitrogen without limit. Instead, they probably only fix just enough nitrogen to allow them to utilize all of the P dissolved in the ocean. In other words, they continue to fix nitrogen as long as there is P available; and once all the P has been consumed in photosynthesis, they stop nitrogen fixation. Over long periods of time, with continuous remineralization of the organic matter produced in photosynthesis, the N:P ratio of the ocean will tend to mimic the N:P ratio in the biosphere itself. If correct, this theory implies that P, rather than N, ultimately limits ocean productivity.

5.3.2. Coupling of the P Cycle to the Rock Cycle—Sedimentation and Weathering

The flow of P into the lithosphere is accomplished through the formation of phosphate-containing sediments at the ocean bottom. The flow of P out of the lithosphere occurs when these same phosphate-containing sediments are brought to the earth's surface at the end of the rock cycle, where they are weathered or eroded. For simplicity, we will treat these processes as a single stoichiometric reaction involving the formation and degradation of hydroxyapatite:

$$5Ca^{2+} + 3HPO_4^{2-} + 4HCO_3^- \rightleftarrows (Ca_5(PO_4)_3OH)_s + 4CO_2 + 3H_2O \qquad (R5.4)$$

Sedimentation proceeds from left to right, and weathering and erosion proceeds from right to left. An important characteristic of this reaction, which is common to many of the sedimentation/weathering reactions in the earth system, is its coupling to C cycle. The production of a mole of hydroxyapatite in sedimentation effectively liberates 4 moles of CO_2 to the atmosphere. On the other hand, the weathering of a mole of apatite requires 4 moles of CO_2 from the atmosphere.

5.4. THE CYCLE

We are now ready to construct our first global biogeochemical cycle—the cycle for P. In doing this it is important to bear in mind that there is no single correct representation of this or any other cycle. It is ultimately the specific issues that you wish to address that will determine the level of detail and sophistication your model will require. In our treatment here, we will adopt a model that includes six boxes or reservoirs. As illustrated in Figure 5.5, these reservoirs are (1) the sediments, (2) the terrestrial soils, (3) the land or terrestrial biota, (4) the ocean biota, (5) the surface ocean, and (6) the deep ocean.[1]

A few aspects of the model presented in Figure 5.5 are worthy of note. P is the tenth most abundant element on the earth, with a relative abundance by weight of about 0.1%. Given that the mass of the earth is about 6×10^{15} Tg (see Table 3.1), we can infer that there is about 6×10^{12} Tg of P on the earth.[2]

---

[1]The six-box model for the P cycle is fairly standard and can be found in a number of other treatments. A few examples are given in the references listed in "Suggested Reading" at the end of the chapter.

[2]Recall that 1 Tg = 1 teragram = 1×10^{12} g.

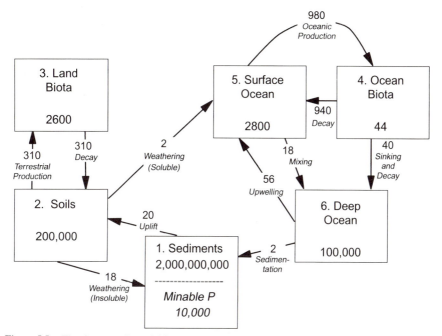

Figure 5.5. The six-reservoir model for the global biogeochemical cycle of P. Reservoirs are given in units of Tg and fluxes are given in units of Tg year^{-1}. (Recall that 1 Tg = 1 × 10^{12} g.)

However, the vast majority of this P resides in the mantle and core and thus does not really participate in the global biogeochemical cycle of P. The P we include in our model is only that portion of the 6 × 10^{12} Tg of P on the earth that is directly involved in the biogeochemical cycle of the element. As indicated in Figure 5.5, this amounts to only about 2 × 10^9 Tg and includes the P found in the (ocean) sediments, the soils, the ocean, and the biosphere. Moreover, because of the special role of oceanic biota as a source of organic sediments at the ocean bottom, we distinguish between the oceanic biosphere and the terrestrial biosphere. Because the exchange of nutrients between the deep ocean and the surface ocean controls the rate of ocean photosynthesis, we treat these two ocean regimes separately. Finally, in order to examine the effect of phosphate mining, we have identified a sub-reservoir within the sediments called "minable P."

Now that we have identified the number and identity of the reservoirs that we will use to represent the global P cycle, we must arrive at estimates for the current amounts of P in each reservoir and for the current fluxes of P from one reservoir to another. As will become obvious from the discussion below, where we briefly outline each reservoir and flux, much of the information we need to make these estimates has already been presented in our summary of the earth system's current metabolic state (Table 3.8).

5.4.1. C_1: The Sediment Reservoir

The total amount of P within sediments, C_1, is estimated at 2×10^9 Tg. This value is derived from our earlier estimate for the total mass of sediments of 2×10^{12} Tg (see Table 3.8) and the average relative abundance of P in the crust of about 0.1% by weight:

$$C_1 = (2 \times 10^{12} \text{ Tg of sediment}) \cdot \left(0.001 \ \frac{\text{g P}}{\text{g sediment}}\right)$$

$$= 2 \times 10^9 \text{ Tg P} \tag{5.4.1}$$

The value of 10^4 Tg of minable P in the sediments is based on the estimates of economic geologists.

5.4.2. C_2: The Terrestrial Soil Reservoir

The P reservoir in terrestrial soils is used to represent that portion of the P on the land that is available for uptake by green plants on the continents. This reservoir amount is derived from our earlier estimates for the total mass of soils in the continents of 2×10^8 Tg and the average P content of the crust of 0.1% by weight. Thus, we find that C_2 is about 2×10^5 Tg P.

5.4.3. C_3: Land Biota Reservoir

The land biota reservoir is estimated from the data we derived in Chapter 3 on the terrestrial biosphere. Recall from Table 3.8 that the size of the living, terrestrial biosphere is estimated at about 8.3×10^5 Tg C and the average P:C ratio in terrestrial organisms is 1:830. Thus,

$$C_3 = (8.3 \times 10^5 \text{ Tg C}) \left(\frac{1 \text{ mole P}}{830 \text{ moles C}}\right) \left(\frac{31 \ \frac{\text{g}}{\text{mole}} \ \text{P}}{12 \ \frac{\text{g}}{\text{mole}} \ \text{C}}\right) \approx 2.6 \times 10^3 \text{ Tg P} \tag{5.4.2}$$

5.4.4. C_4: Ocean Biota Reservoir

Similar to our estimate for the land biota reservoir, the ocean biota reservoir is based on the size of the living, oceanic biosphere of 1.8×10^3 Tg C and the average P:C ratio in marine organisms of 1:106 (see Table 3.8). Thus,

$$C_4 = (1.8 \times 10^3 \text{ Tg C}) \left(\frac{1 \text{ mole P}}{106 \text{ moles C}}\right) \left(\frac{31 \ \frac{\text{g}}{\text{mole}} \ \text{P}}{12 \ \frac{\text{g}}{\text{mole}} \ \text{C}}\right) \approx 4.4 \times 10^1 \text{ Tg P} \tag{5.4.3}$$

5.4.5. C_5: Surface Ocean Reservoir

The P reservoir in the surface ocean is obtained from the average dissolved P concentration in surface waters of about 25 mg m^{-3}, the mass of the surface ocean of 1.1×10^{11} Tg, and an ocean density of 1 g cm^{-3}. Combining these data, we obtain a value of 2.8×10^3 Tg P for C_5.

5.4.6. C_6: Deep Ocean Reservoir

The deep ocean reservoir is similarly estimated from (a) an average P concentration of 80 mg m^{-3} in the deep ocean and (b) a deep ocean mass of 1.3×10^{12} Tg. The resulting estimate for C_6 is 1.0×10^5 Tg P.

5.4.7. $F_{2 \to 1}$: Flux from Land to Sediment

$F_{2 \to 1}$ is used to represent the effective flux from the land to the sediment that results from the mechanical weathering of insoluble phosphate minerals. The weathering of this insoluble phosphate produces particles or silt that are carried by the rivers into the ocean, where they fall to the ocean bottom and become part of the ocean sediments. Because the phosphorus in these particles never dissolves into the ocean water, we can treat this transport as a direct transfer of phosphorus from the land to the sediments. The magnitude of $F_{2 \to 1}$ is computed from (a) the estimated weathering rate of 2×10^4 Tg per year, (b) the average P content of the crust of 0.1% by weight, and (c) an estimate that 90% of the total weathering produces insoluble phosphorus. The resulting flux is thus given by

$$F_{2 \to 1} = \left(2 \times 10^4 \; \frac{\text{Tg sediment}}{\text{year}} \right) \left(0.001 \; \frac{\text{g P}}{\text{g sediment}} \right) (0.9) = 18 \; \frac{\text{Tg P}}{\text{year}}$$

(5.4.4)

5.4.8. $F_{2 \to 3}$: Flux from Land to Land Biota

This flux is computed from the estimated gross primary production rate for the terrestrial biosphere of 1×10^5 Tg C per year and the average P:C ratio in land biota of 1:830. The resulting flux is 310 Tg P year^{-1}.

5.4.9. $F_{3 \to 2}$: Flux from Land Biota to Land

The return flux from land biota to the land is assumed to be equal to $F_{2 \to 3}$ (i.e., 310 Tg P year^{-1}), yielding a balance in the production and loss of P in the terrestrial biosphere.

5.4.10. $F_{2 \to 5}$: Flux from Land to Surface Ocean

The estimate of this flux is based on the same data as those used for $F_{2 \to 1}$, only here we assume that 10% of the weathering of phosphate minerals leads to dissolved phosphorus that is carried by river runoff into the ocean as orthophosphate. Thus, $F_{2 \to 6} = 2$ Tg P year^{-1}.

5.411. $F_{5 \to 4}$: Flux from Surface Ocean to Ocean Biota

The uptake of P in the surface ocean by ocean biota is estimated from the gross surface ocean productivity of 4×10^4 Tg C year^{-1} and the average P:C ratio in marine organisms of 1:106. The resulting flux is given by

$$F_{5 \to 4} = \left(4 \times 10^4 \frac{\text{Tg C}}{\text{year}}\right)\left(\frac{1 \text{ mole P}}{106 \text{ moles C}}\right)\left(\frac{31 \frac{\text{g}}{\text{mole}} \text{ P}}{12 \frac{\text{g}}{\text{mole}} \text{ C}}\right)$$

$$\approx 9.8 \times 10^2 \frac{\text{Tg P}}{\text{year}}$$

$$(5.4.5)$$

5.4.12. $F_{4 \to 5}$: Flux from Ocean Biota to Surface Ocean

$F_{4 \to 5}$ represents the return of P from ocean biota to the surface ocean as a result of respiration and decay within the surface ocean. It is estimated that roughly 96% of the P assimilated by phytoplankton in gross surface ocean production is immediately remineralized in the surface ocean. (Recall from Chapter 3 that only about 90% of the C in gross surface ocean productivity is returned to the surface ocean. However, P is generally remineralized more efficiently during the decay of organic material than is C; hence, more of it is returned to the surface ocean and less is lost via sinking to the deep ocean.) Thus

$$F_{4 \to 5} = 0.96 \, F_{5 \to 4} = 9.4 \times 10^2 \frac{\text{Tg P}}{\text{year}} \qquad (5.4.6)$$

5.4.13. $F_{4 \to 6}$: Flux from Ocean Biota to Deep Ocean

The flux of P from ocean biota to the deep ocean is the 4% of the P assimilated by phytoplankton that is not remineralized in the surface ocean. It follows therefore that $F_{4 \to 6} = 40$ Tg P year^{-1}.

5.4.14. $F_{5 \to 6}$: Flux from Surface Ocean to Deep Ocean

The rate of exchange of inorganic P from the surface ocean to the deep ocean is derived from the surface ocean/deep ocean exchange coefficient of 2 m year^{-1}, the concentration of P in the surface ocean of 25 mg m^{-3}, and the surface area of the ocean of 3.5×10^{18} cm^2:

$$F_{5 \to 6} = \left(25 \frac{\text{mg P}}{\text{m}^3}\right)\left(10^{-3} \frac{\text{g}}{\text{mg}}\right)\left(10^{-6} \frac{\text{m}^3}{\text{cm}^3}\right)\left(200 \frac{\text{cm}}{\text{year}}\right)(3.5 \times 10^{18} \text{ cm}^2)$$

$$\approx 1.8 \times 10^1 \frac{\text{Tg P}}{\text{year}}$$

$$(5.4.7)$$

5.4.15. $F_{6 \to 5}$: Flux from Deep Ocean to Surface Ocean

The rate of exchange of P from the deep ocean to the surface ocean is calculated using the same exchange coefficient of 2 m year^{-1} with a deep ocean P concentration of 80 mg m^{-3}. The resulting flux is 56 Tg P year^{-1}.

5.4.16. $F_{6 \to 12}$: Flux from Deep Ocean to Sediments

The rate at which P accumulates in the ocean sediments via precipitation of apatite and other phosphate minerals is a complicated function of the phosphate concentration in the deep ocean and a host of other variables that can affect the rate at which phosphate crystals can form in the ocean. For simplicity, we will specify this flux by invoking the contemporary steady-state assumption; that is, we will assume that the rate of phosphate sedimentation is exactly equal to the amount needed to create a balance between the input of phosphorus to the deep ocean and the loss of phosphorus from the deep ocean. With this assumption, $F_{6 \to 1} = 2$ Tg P year^{-1}.

5.4.17. $F_{1 \to 2}$: Flux from Sediments to Soils

As we did in the case of $F_{6 \to 1}$, we will specify the flux of P from the sediments to the continents by invoking steady-state. Thus, we assume that $F_{1 \to 2} = 20$ Tg P year^{-1}.

5.4.18. Setting Up the K Matrix

We have now specified all the reservoir amounts and fluxes in our cycle. The last task we must complete before we can actually carry out numerical experiments in our cycle is the specification of the exchange coefficients or k values. Recalling from Chapter 4 that

$$k_{i \to j} = \frac{F_{i \to j}}{C_i} \tag{5.4.8}$$

and that each vertical column of the **K** matrix is composed of the $k_{i \to j}$ elements for a specific value of i and j varying from 1 to N, it is a straightforward task to show that

$$
\mathbf{K} =
\begin{array}{cccccc}
-1 \times 10^{-8} & 9 \times 10^{-5} & 0.0 & 0.0 & 0.0 & 2 \times 10^{-5} \\
1 \times 10^{-8} & -1.65 \times 10^{-3} & 0.119 & 0.0 & 0.0 & 0.0 \\
0.0 & 1.55 \times 10^{-3} & -0.119 & 0.0 & 0.0 & 0.0 \\
0.0 & 0.0 & 0.0 & -22.27 & 0.35 & 0.0 \\
0.0 & 1.0 \times 10^{-5} & 0.0 & 21.36 & -0.35643 & 5.6 \times 10^{-4} \\
0.0 & 0.0 & 0.0 & 0.91 & 6.43 \times 10^{-3} & -5.8 \times 10^{-4}
\end{array}
$$

In the next section we will use the numerical techniques developed in Chapter 4 to explore the P cycle in more detail.

5.5. EXPLORING THE CYCLE WITH *BOXES*

We are now in a position to use *Boxes*, or any other equivalent mathematical technique, to explore the inner workings of the P cycle. There are, in fact, any number of different numerical experiments that one could carry out. A few instructive examples are presented below.

5.5.1. Experiment 1: Verifying the Steady-State Model

Once you have set up a biogeochemical cycle and specified its **K** matrix, it is always a good idea to make sure that you have done it correctly. If the cycle you have set up is in steady state, this is easily done using *BOXES*: Simply input the **K** matrix and specify the initial amounts for each reservoir to be the amounts that give a steady-state cycle. Then use the menu in *BOXES* to "Run/Solve." Because the cycle is in steady state, you should see no change if you have done everything correctly (see Text Box 5.2).

In addition to verifying your model, the steady-state solution can provide you with some information on the characteristics of the cycle if you are willing to dig a little deeper. To do this, go into *BOXES* and view the results in text format using the "Output/Screen" command. Scroll down to the section labeled "Eigenvalue Vector." There you will find a listing of the eigenvalues for our cycle. For the six-box model of the phosphorus cycle, you should find five nonzero eigenvalues and 1 eigenvalue of zero (see Figure 5.7). Note that the eigenvalue with the smallest mag-

TEXT BOX 5.2	USING OUTPUT FROM *BOXES* TO VERIFY THE STEADY-STATE MODEL

Once you have set up your steady-state cycle, you can use *BOXES* to make sure that you have done everything correctly. Input the steady state reservoir amounts and *k* values as instructed in the Help menus and then use your mouse to activate the "Run/Solve" command. Once the calculations are complete, you can check to make sure your model is in a steady state in either of two ways. The "Plot/View" command will produce a graphical representation of the reservoir amounts as a function of time on your computer screen. If you have calculated the *k* values correctly and properly input the data, the plot should be nothing more than a series of straight lines showing no change with time as in Figure 5.6. Alternatively, you can take a look at the output in text format using the "Output/Screen" command. When the output appears on the computer screen, you will find several different sections of numbers with different labels. The first three sections provide you with a record of the reservoir names, their initial amounts and the **K** matrix you used to run the model. The rest of the sections document the solution. Verifying that we have input the steady-state model can be obtained by tabbing down to the last two lines of the file. The next to the last line is the one labeled "Infinite." This is where *BOXES* lists reservoir amounts in each reservoir at $t = \infty$—that is, when all the transient parts of the solution have decayed away. The last line is labeled "Change" and is where *BOXES* lists the net change in each reservoir amount from $t = 0$ to $t = \infty$. If you have correctly input the steady-state model, the "Infinite" solutions should be essentially identical to the initial reservoir amounts, and the "Changes" should be only a few percent or less of the initial and final reservoir amounts. Do you know why the "Changes" are not exactly equal to zero?

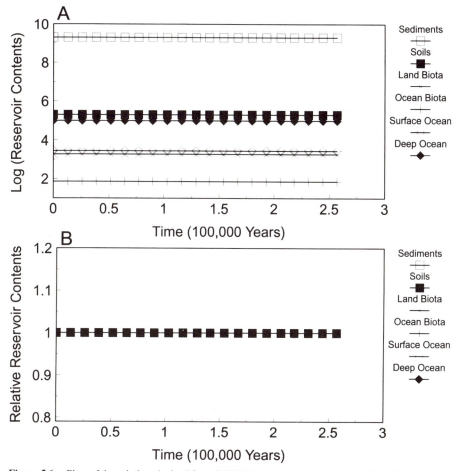

Figure 5.6. Plots of the solution obtained from *BOXES* for the steady-state P cycle. Plots like the ones illustrated here can be obtained with *BOXES* using the "Plot/Options" and "Plot/View" commands to select and view. (**A**) A plot of the contents of each reservoir. (**B**) A plot of the relative reservoir contents— that is, the content of each reservoir divided by its initial content. (Alternatively, users can use the "Output/File" command to save the results in text format to a file and then import this file into an appropriate spreadsheet or graphics application.) Regardless of the option adopted, a plot of the results from a steady-state cycle should always show constant reservoir amounts as a function of time, as we obtain here for the steady-state P cycle.

nitude is 1.9×10^{-5} year^{-1}. This eigenvalue corresponds to the inverse of the characteristic response (or *e*-folding) time of the cycle. In other words, if we should perturb the cycle in any way, about 50,000 years (i.e., $1 \div 1.9 \times 10^{-5}$ year^{-1}) will be required for all of the reservoirs to adjust to this perturbation and approach a new steady state. You might find this result a little surprising. Recall that the residence for P in the sediments is about 200 million years (i.e., $1 \div k_{1 \rightarrow 2}$). How is it possible that the response time for the entire system can be only 50,000 years when the residence time for P in one of the reservoirs is 200 million years?

Eigenvalue Vector:

$-1.376E+01$ $-1.646E-01$ $-2.077E-02$ $-9.907E-05$ $-1.946E-05$ $0.000E+00$

Figure 5.7. The eigenvalues for the P cycle, as listed in the output file from *BOXES*. The last eigenvalue corresponds to the steady-state solution, and the others correspond to the five transient solutions. The inverse of the next to the last eigenvalue, about 50,000 years, is the maximum time required by the cycle to respond to a perturbation and re-equilibrate. This time is referred to as the characteristic response time of the cycle.

5.5.2. Experiment 2: Impact of Anthropogenic Activities

As indicated schematically in Figure 5.5, the mining of phosphates and their use in agriculture as fertilizer and a variety of other industrial processes have the effect of accelerating the rate at which P in the sediments is transferred to the soils and ultimately made available to the biota. To investigate the long-term impact of this anthropogenic perturbation of the P cycle, we adopt a "worst-case scenario": Suppose that all the minable P in the sediments, estimated at 10,000 Tg, was suddenly moved from the sediments to the soils reservoir. What would be the effect on the P cycle?

This problem can be easily addressed using *BOXES*, by making a small change to the input file we constructed to simulate the steady-state cycle in Section 5.5.1. We simply add 10,000 Tg to the initial content of the soils reservoir (i.e., reservoir #2). Note that because we have not made any alterations to the mechanisms that control the exchange of P from reservoir to another, the **K** matrix is not altered.

As illustrated in Figure 5.8, the resulting perturbations to the P cycle are not very large. In the short term (i.e., time scales of decades to centuries), the initial 5% increase in the reservoir of P in the soils leads to a corresponding enhancement in the mass of land biota. Eventually however, the excess P in the soils and land biota begins to drain into the ocean, leading, in turn, to a small enhancement in oceanic P. This enhancement in oceanic P also decays, as the excess P ultimately finds itself back in the sediments. After about 50,000 years, the cycle has re-equilibrated with much the same amounts of P in each of the reservoirs as there was before the minable P was moved out of the sediments.

The relatively minor effects produced in our global model from the dumping of all minable P in the soils should not be interpreted to mean that anthropogenic perturbations to the P cycle have not had significant environmental consequences. In fact, the additions of anthropogenic phosphates to rivers and streams have caused major perturbations to the nutrient cycles of any number of fresh water systems and even the complete eutrophication of some lakes. However, these effects have been largely limited to local areas strongly impacted by anthropogenic activities. In our global model, we are forced to make the simplifying and incorrect assumption that the P is added uniformly to all soils, which effectively dilutes the impact of the anthropogenic perturbation. Thus we find an important limitation to the approach we have adopted here: Because we take a global perspective, we are not able to catch the regional heterogeneity that characterizes many important aspects of biogeochemical cycles.

5.5.3. Experiment 3—Doubling Photosynthesis

In our third experiment, we consider what might happen if the rates of photosynthesis by both the terrestrial and oceanic biota doubled (perhaps as a result of an increase in atmospheric CO_2). In contrast to the previous experiment, which necessitated a change in the initial reservoir contents, this perturbation requires that we change the **K** matrix. Specifically, we must (i) double the values of $k_{2\to3}$ and $k_{5\to4}$ that control the transfer of P from the soils and surface ocean to the land and ocean biota, respectively, and (ii) adjust the diagonal **K**-matrix elements K_{22} and K_{55} that relate to the total flow of P out of reservoirs 2 and 5, respectively. The resulting **K** matrix takes the following form:

$$\mathbf{K} = \begin{bmatrix} -1 \times 10^{-8} & 9 \times 10^{-5} & 0.0 & 0.0 & 0.0 & 2 \times 10^{-5} \\ 1 \times 10^{-8} & \mathbf{-3.2 \times 10^{-3}} & 0.119 & 0.0 & 0.0 & 0.0 \\ 0.0 & \mathbf{3.1 \times 10^{-3}} & -0.119 & 0.0 & 0.0 & 0.0 \\ 0.0 & 0.0 & 0.0 & -22.27 & \mathbf{0.70} & 0.0 \\ 0.0 & 1.0 \times 10^{-5} & 0.0 & 21.36 & \mathbf{-0.70643} & 5.6 \times 10^{-4} \\ 0.0 & 0.0 & 0.0 & 0.91 & 6.43 \times 10^{-3} & -5.8 \times 10^{-4} \end{bmatrix}$$

where bold numbers indicate those elements of the matrix that have been changed. (Note that because this perturbation does not involve an initial change in the amount of P in the reservoirs, the initial reservoir contents input into *BOXES* for this experiment should be the same as those input for the steady-state simulation.)

The response of the land biota, ocean biota, and surface ocean reservoirs to this perturbation is illustrated in Figure 5.9. The change in the land biota reservoir is relatively straightforward, with a monotonic increase in the P content from its ini-

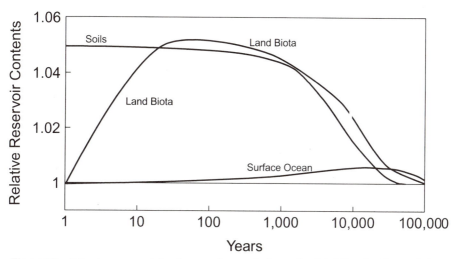

Figure 5.8. "Worst-case scenario" anthropogenic perturbation to the global P cycle. Time variations in the relative amounts of P in the soils, land biota, and surface ocean reservoirs after a sudden transfer of the entire reserve of minable P from the sediments to the soils at time $t = 0$. By "relative contents" we mean the contents of each reservoir relative to its steady state amount, as calculated in Experiment 1.

tial value of 2600 Tg to a final content of about 5200 Tg. The time constant for this change is relatively short (about 10 years), and thus the new equilibrium is essentially reached after only a few decades. By comparison, the variation in the ocean biota is more complex, with an initial rise and then a drop in P content. The final ocean biota reservoir, also reached within 20 to 30 years, is only 52 Tg P, a net increase of only 8 Tg from its initial content of 44 Tg. Why the rather small increase in ocean biota in spite of a doubling in the photosynthetic rate? The answer can be found in the variation in the surface ocean reservoir, which decreases from an initial value of 2800 Tg to a final amount of about 1660 Tg.

Our model has produced an interesting, and what might at first seem to be a surprising and nonintuitive, result: A doubling in the rate of photosynthesis actually leads to a depletion of P from the surface ocean and only a ~20% increase in the oceanic biosphere . The reason for this can be seen by looking again at our schematic of the P cycle in Figure 5.5. The primary pathway for transporting P from the sur-

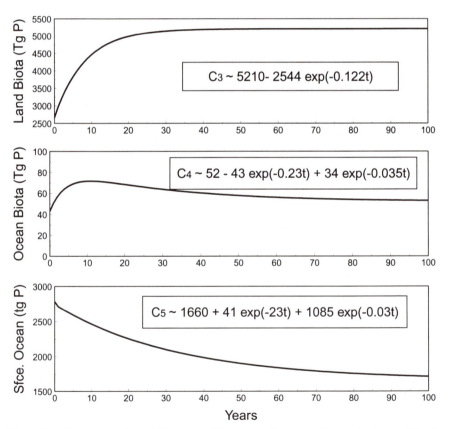

Figure 5.9. Response of the land biota, ocean biota, and surface ocean P reservoirs to a doubling in the rates of photosynthesis. The functions within each box depict the approximate analytical form of the solution for each of the respective reservoirs. These functional forms can be viewed in *BOXES* using the "Output/Screen" command and tabbing down to the "General Solution" section.

face to the deep ocean is through the sinking and decay of ocean biota (i.e., the nutrient cycle discussed in Chapter 3). By increasing photosynthesis, we shift the balance of P in the surface ocean between inorganic and organic forms of P and thus enhance the rate of transfer of P from the surface to the deep ocean. The result is a depletion of P from the surface ocean.

Thus we find one of the most significant benefits of the numerical approach we have adopted here: the ability to uncover feedbacks and interactions within a biogeochemical cycle that lead to unanticipated results and insights into the cycle. Two other numerical experiments that might provide the reader with additional insights into the P cycle are included in the problems at the end of the chapter.

5.6. CONCLUSION

Because P is limited to its +5 oxidation state in the earth system, its biogeochemical cycle is relatively simple and, thus, not hard to construct. Nevertheless, the analysis of this cycle has provided us with some important lessons. One of these lessons relates to the limitations of the global perspective we have adopted here. By definition, a global approach ignores regional and local inhomogeneities and thus perturbations that are important on these smaller scales (such as that caused by the mining of P and its eventual deposition on soils) may be lost on the global scale.

Another important lesson is that related to the complexity of the earth system and the processes at work in a global biogeochemical cycle. Because of this complexity, a biogeochemical cycle can sometimes respond to a perturbation (such as a doubling in the oceanic photosynthetic rate) in nonlinear and, in some cases, even counterintuitive ways. For this reason, a mathematical description of biogeochemical cycles, such as that in *BOXES*, can be an indispensable diagnostic tool for the biogeochemist.

In the next chapter we turn to a more complex biogeochemical cycle—that involving the element carbon.

SUGGESTED READING

Lerman et al., *Geological Society of America*, **142**, 205–218, 1975.

Garrels, R. M., F. T. MacKenzie, and C. Hunt, *Chemical Cycles and the Global Environment*, W. Kaufman, Los Altos, California, 1975.

Graham and Duce, *Geochimica et Cosmochimica Acta*, **43**, 1195–1208, 1979.

Jahnke, R. A., The phosphorus cycle, in *Global Biogeochemical Cycles*, edited by S. S. Butcher, R. J. Charlson, G. H. Orians, and G. V. Wolfe, Acadmic Press, New York, 301–315, 1992.

Pierrou, U., The global phosphorus cycle, in *Nitrogen, Phosphorus, and Sulfur—Global Cycles*, Scientific Committee on Problems of the Environment (SCOPE) Report 7, edited by B. H. Svensson and R. Soderlund, Royal Swedish Academy of Sciences, pp. 75–88, 1975.

Van Cappellen, P., and E. Ingall, Redox stabilization of the atmosphere and oceans by phosphorus-limited marine productivity, *Science*, **271**, 493–496, 1996.

PROBLEMS

1. Why is the response time for the P cycle only 50,000 years when the residence time for P in the sediments is 200 million years?

2. The "Doomsday Scenario": Imagine all photosynthesis suddenly stopped. Use *BOXES* to predict what the short- and long-term response of the P cycle would be to this catastrophic event.

3. Simulating an Ice Age: The past ~2 million years has been characterized by a fluctuating climate with glacial periods (i.e., ice ages) alternating with interglacial periods every 100,000 years or so. During the ice ages it is believed that the ocean circulation, along with its mixing of surface and deep ocean waters, is significantly depressed. Use the model for the P cycle developed in this chapter to investigate the effect of an ice age on the terrestrial and oceanic biota and the recovery of same after the return to an interglacial. *Hint:* Start with the steady-state cycle. Assume that at time $t = 0$ a 50,000-year ice age begins and during this period there is no ocean mixing. The absence of ocean mixing will require that you adjust the appropriate k values that relate to mixing of P between the surface and deep ocean. Run *BOXES* for 50,000 years. Then reinstate your initial k values and run the system for another 50,000 years.

The Global Carbon Cycle

6

> "I am, reluctantly, a self-confessed carbon chauvinist. Carbon is abundant in the Cosmos. It makes marvelously complex molecules, good for life... But I sometimes wonder. Could my fondness for these materials haves something to do with the fact that I am made chiefly of them?"
>
> C. Sagan, *Cosmos*, Random House, New York, 1980.

6.1. INTRODUCTION

Carbon's unassuming position in the Periodic Table, first row in Group IVA (Figure 6.1), belies its central and unique role in chemistry and biology. This element's singular role arises from its ability to form strong, stable bonds with other carbon (C) atoms. In fact, the C–C bond tends to be as strong as the bonds that C forms with many other elements; as a result, there are an almost limitless number of chained and ringed molecules with C atoms bonded to other C atoms; these molecules, when they occur in conjunction with hydrogen, oxygen, nitrogen, and phosphorus, comprise the stuff of life. For this reason the biogeochemical cycle of C is central to our understanding of the earth system.

However, the desire to understand the global carbon cycle arises from more than just the intellectual curiosity of earth system scientists. The concentration of atmospheric carbon dioxide has been steadily increasing since the Industrial Revolution (Figure 6.2). There can be no doubt that this increase has been caused by an increasingly populous and technological human society. Anthropogenic sources of carbon dioxide include the mining and burning of fossil fuels, the manufacturing of cements, and the clearing of forests. Because of carbon dioxide's radiative properties, it acts as a *greenhouse gas* that acts to warm the earth's surface. It is feared that a global climate warming will eventually ensue, if it has not already begun, if the atmospheric concentration of this gas continues to increase (Text Box 6.1). Key to being able to predict how large global temperatures might warm is being able to predict how much carbon dioxide concentrations will increase in the coming decades and how long this increase will per-

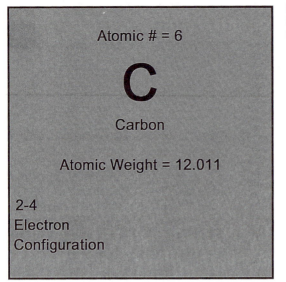

Figure 6.1. Carbon is a nonmetallic element in Group IVA of the Periodic Table.

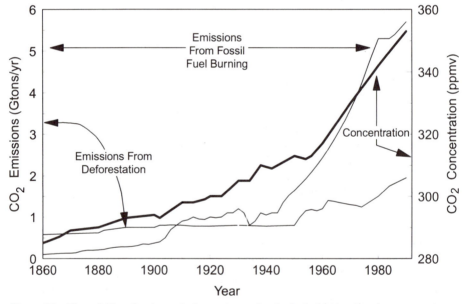

Figure 6.2. The activities of an increasingly populous and technological society have become a major force in the global cycle of C, causing a dramatic increase in the atmospheric concentrations carbon dioxide (CO_2) since the Industrial Revolution. Illustrated here is the close correlation since 1860 between the annual emissions from fossil fuel burning and deforestation and the annually averaged CO_2 concentration, as inferred from analyses of air trapped in ice and by direct atmospheric measurements. Because of CO_2's role as a greenhouse gas, this increase has caused concerns of an imminent global warming. Central to assessing the possibility and severity of this warming is knowing how long the excess CO_2 placed in the atmosphere by humankind will persist. In this chapter we will use *BOXES* to address this question. (Figure prepared from data in Intergovernmental Panel on Climate Change, *Climate Change 1994: Radiative Forcing of Climate Change*, Cambridge University Press, New York, 1995 and references cited therein.)

| TEXT BOX 6.1 | CO$_2$ AND AN ENHANCED GREENHOUSE EFFECT |

The vast majority of the energy from the sun comes to earth in the form of visible light. Because the atmosphere is mostly transparent to radiation in the visible portion of the spectrum, most of the energy from the sun passes right through the atmosphere and is absorbed at the earth's surface. To maintain a balance between incoming and outgoing energy, the earth cools by radiating energy back out to space. However, the earth is considerably cooler than the sun; and, as a result, most of earth's energy is radiated in the infrared portion of the spectrum. It turns out that the atmosphere contains a number of gases—the so-called *greenhouse gases*—that are transparent to visible radiation but opaque to infrared radiation. These gases thus absorb radiation from the earth's surface and re-radiate a portion of it back to the earth's surface. This causes the earth's surface to heat up, bringing about the *greenhouse effect* (see Figure 6.3).

The current debate on global warming is not about whether or not a greenhouse effect exists, because it clearly does. If the atmosphere did not contain any greenhouse gases, temperatures at the earth's surface would be some 20°C below the freezing point of water and the earth would be a very different planet, perhaps even devoid of life. Nor is the debate about whether the increasing concentrations of CO$_2$ and other greenhouse gases will cause an *enhanced greenhouse effect*: Changes in the concentrations of greenhouse gases since the mid-1800s are estimated to have already increased the tropospheric heating rate (or *radiative forcing*) by about 2–3 W m^{-2} (see Figure 6.4).

From a scientific point-of-view, the debate is instead focused on the magnitude of the climatic perturbation that will result from this radiative forcing. The sensitivity of the global climate system to a given radiative forcing is often expressed in terms of the following simple relationship:

$$\Delta T_s = \lambda \Delta F$$

where ΔT_s is the change in the global mean surface temperature (in K) due to a radiative forcing of ΔF (in W m^{-2}), and λ is the so-called *climate sensitivity parameter*. Depending upon how climate models treat a number of poorly understood phenomena such as the hydrologic cycle, λ is found to vary from a few tenths to a few °K per W/m^{-2}. At the low end of this range, the impact of the enhanced greenhouse effect will probably be relatively minor, but at the high end the impact will probably be quite severe.

From a public policy point of view, the debate is focused on what to do in the face of this scientific uncertainty. Some argue that we should do nothing until we can be certain that the climatic perturbation will be significant, while others argue that even the possibility of a significant climate warming demands immediate action to curb the buildup of greenhouse gases. One critical aspect of this debate is the reversibility of the climatic perturbation: If we do nothing now and allow greenhouse gases to continue to increase in abundance and a climatic perturbation does occur, will the atmosphere rid itself of the excess greenhouse gases and return to its original state if we then halt further greenhouse gas emissions? And, if so, how long will it take? In this chapter, we will address a key facet of this question, namely, the lifetime of the excess carbon dioxide.

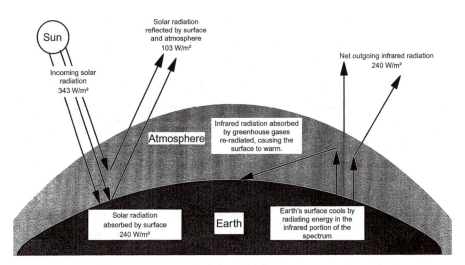

Figure 6.3. A simplified illustration of the earth's radiative balance and the role of atmospheric green-house gases. Roughly two-thirds of the incoming solar radiation passes through the atmosphere and is absorbed at earth's surface. The earth establishes an energy balance by radiating infrared radiation back to space. Atmospheric greenhouse gases absorb a portion of the infrared radiation from the earth's sur-face and, in turn, re-radiate a portion of it back to the surface, thereby creating the greenhouse effect. In the absence of the greenhouse effect, the average surface temperature of the earth would about 255 K, some 33 K below its present value. (After Intergovernmental Panel on Climate Change, *Climate Change 1994: Radiative Forcing of Change*, Cambridge University Press, New York, 1995.)

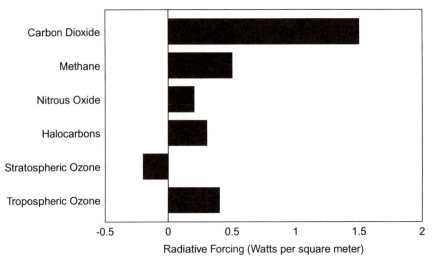

Figure 6.4. Estimated radiative forcing (i.e., net radiative heating of the troposphere) caused by changes in the concentrations of carbon dioxide, methane, nitrous oxide, halocarbons, stratospheric ozone, and tropospheric ozone since the mid-1800s. (After Intergovernmental Panel on Climate Change, *Climate Change 1994: Radiative Forcing of Climate Change*, Cambridge University Press, New York, 1995.)

sist. In this chapter we will see how an understanding of the global biogeochemical cycle of carbon can help provide answers to this facet of the global warming debate.

6.2. C REDOX CHEMISTRY

As we did in the previous chapter, we begin our analysis of the C cycle with a brief review of the element's thermochemistry. From Figure 4.1, we see that C has four electrons in its outer shell and thus it can occupy eight oxidation states ranging from -4 to $+4$. Outside of the biosphere, C is most often found in its two most extreme oxidation states: (i) the $+4$ oxidation state as gaseous carbon dioxide (CO_2) in the atmosphere, carbonic acid and its conjugate bases (i.e., H_2CO_3, HCO_3^-, and CO_3^{2-}) in the hydrosphere, and carbonate minerals such as $CaCO_3$ in the lithosphere; and (ii) The -4 oxidation state as gaseous methane (CH_4) in the atmosphere and *methane clathrates* in the lithosphere. C is also found in its intermediate oxidation states—for instance, in the $+2$ oxidation state as carbon monoxide (CO), although generally in much less abundance than the $+4$ and -4 oxidation states. Organic C in the biosphere can also occupy a variety of oxidation states, ranging from $+2$ in esters (i.e., RC(O)OC) to $-2(n+1/n)$ in alkanes (i.e, C_nH_{2n+2}).

For simplicity, we will consider three possible oxidation states for C in our analysis of C's redox chemistry: $+4$, 0, and -4 with the 0 oxidation state, appropriate for C in glucose ($C_6H_{12}O_6$), thermodynamically equivalent to "CH_2O" our so-called generic organic carbon molecule. The equilibria between these three oxidation states are summarized in Figure 6.5. Following the methodology described in Chapter 2 and using the data summarized in this figure and in the Appendix, we can derive a series of linear equations that describe the stability boundaries between these three oxidation states in pe–pH phase space. Assuming equilibrium partial pressures of 1 atm for CO_2 and CH_4, these equations are as follows:

1. For the boundary between CH_4 and CO_2 we have

$$pe = 2.86 - pH \qquad (6.1)$$

2. For the boundary between $C_6H_{12}O_6$ and CH_4 we have

$$pe = 6.81 - pH \qquad (6.2)$$

3. For the boundary between $C_6H_{12}O_6$ and CO_2 we have

$$pe = -1.08 - pH \qquad (6.3)$$

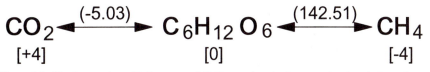

Figure 6.5. The three major oxidation states of C. The numbers in the square brackets indicate the oxidation states of C within the compounds, and the numbers in the parentheses the log of the equilibrium constant for the redox half-reaction between the two compounds.

These lines, along with the stability field of water, are plotted on a pe–pH diagram in Figure 6.6.

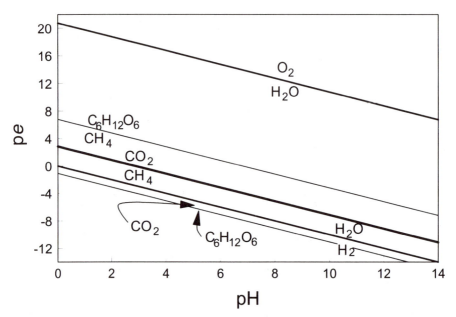

Figure 6.6. The pe–pH stability diagram for C superimposed upon the stability field for water. Stability boundaries for water are indicated by the solid lines, and metastable boundaries for C are indicated by the thin lines. The stability boundary between CO_2 and CH_4 is indicated by the heavy solid line. Note that at any combination of pe and pH, $C_6H_{12}O_6$ is unstable with respect to either CO_2 or CH_4. Above the CO_2–CH_4 boundary (i.e., at high pe), CO_2 is more stable than CH_4 and so the $C_6H_{12}O_6$–CH_4 boundary is a metastable boundary. Similarly, below the CO_2–CH_4 boundary (i.e., at low pe), CH_4 is more stable than CO_2 and thus the $C_6H_{12}O_6$–CO_2 boundary is a metastable boundary.

Inspection of Figure 6.6 reveals that both the CO_2–$C_6H_{12}O_6$ and CH_4–$C_6H_{12}O_6$ boundaries are metastable.[1] We can therefore conclude that only CO_2 and CH_4 are thermodynamically stable within earth system, with CO_2 being stable under more oxidizing conditions and CH_4 being stable under reducing conditions. It follows that organic C, treated here as glucose, is thermodynamically unstable. In fact, although the above calculations were limited to only one organic compound, this conclusion is applicable to all organic molecules regardless of the specific oxidation state of the C atoms. And thus we find that organic compounds, including the chemicals that make up our own protoplasm, are thermodynamically unstable in the earth's environment. For this reason, autotrophs must expend energy to synthesize organic material, and all organisms must constantly expend energy to keep their protoplasm isolated from a thermodynamically hostile environment. Indeed, one of the ironies of the earth system is the fact that the oxygen upon which we respirators so critically depend is actually a toxic substance that would oxidize and destroy our cells if it came in direct contact with them. This tension between an organism's need to be isolated from the earth's oxidizing environment and the very same organism's need to use the oxidants from this environment defines the underlying basis for the biogeochemical cycle of C.

---

[1]Recall from Chapter 2 that a metastable boundary is one between two species where neither represents the most stable form of the element. For example, the boundary between $C_6H_{12}O_6$ and CH_4 lies in a region of pe–pH space where CO_2 is more stable than both $C_6H_{12}O_6$ and CH_4. Hence the boundary is a metastable boundary.

6.3. THE PREINDUSTRIAL GLOBAL BIOGEOCHEMICAL CYCLE OF C

Figure 6.7 presents a schematic illustration of the "preindustrial" global biogeochemical cycle of C. We use "preindustrial" to denote some point in time (say 100–200 years ago), before anthropogenic activities began to cause atmospheric CO_2 concentrations to increase by a significant amount. We adopt an eight-box model for the C cycle, with reservoirs for the atmosphere, the living and dead terrestrial biospheres, the marine biosphere, the surface and deep oceans (comprised primarily of carbonates but also containing dissolved and particulate organic matter), sediments of carbonates (comprised mostly of calcium carbonate), and organic C from which the reserves of fossil fuel are currently mined. The C content of the atmospheric reservoir of 600 Gtons C is derived from Equation (3.9) assuming a preindustrial atmospheric CO_2 mixing ratio of 280 ppmv (see Figure 6.1).[2] The amount of C in each of the biospheric reservoirs is obtained from the data in Table 3.14, and the sizes of the ocean and sediment reservoirs are derived from data on the concentrations of C in each component of the earth system.

Note that, unlike the P cycle discussed in Chapter 5, we use separate reservoirs for the living and dead portions of the terrestrial biosphere (i.e., Reservoirs 2 and 3). On the other hand, dead organic C in the ocean is not included in the ocean bios-

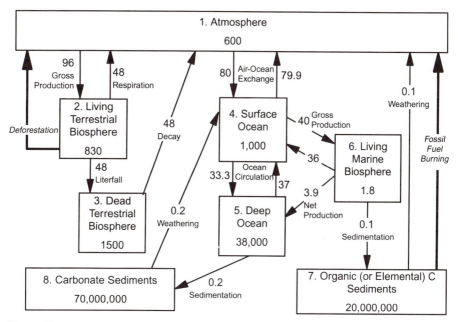

Figure 6.7. The preindustrial, steady-state global cycle for C using an eight-box model. The boxes represent specific reservoirs, with the numbers inside the boxes being reservoir amounts in units of Gtons C and the arrows fluxes between reservoirs with the numbers having units of Gtons C/year. The heavy arrows denote the pathways by which anthropogenic activities perturb the global cycle.

---

[2]1 Gton C = 1×10^9 tons = 1×10^{15} g.

phere but rather placed in the surface and deep ocean C reservoirs (i.e., Reservoirs 4 and 5). The different treatments for the terrestrial and oceanic biospheres are adopted because of the very different pathways organic C takes in the two regimes. In the terrestrial system, both living and dead organic C can be transferred directly to the atmosphere via respiration and decay, and thus separate reservoirs for the kinds of organic C are useful. In the ocean, however, the primary pathways for the loss of organic C is from the living biosphere to the surface ocean via respiration and decay and to the deep ocean via sinking and then oxidation to inorganic form (see Figure 3.14). For this reason, a dead marine biospheric reservoir provides little advantage and is not included in our analysis.

The C cycle is dominated by two sets of biogeochemical reactions. The first set of these reactions includes the processes of photosynthesis

$$CO_2 + H_2O + h\nu \rightarrow \text{``}CH_2O\text{''} + O_2 \tag{R1.1}$$

and respiration and decay

$$\text{``}CH_2O\text{''} + O_2 \rightarrow CO_2 + H_2O \tag{R1.2}$$

that cause of the flow of C into and out of the terrestrial and marine biospheres.[3]

The other set of key biogeochemical reactions includes those that couple the C cycle to the rock cycle through the formation of organic C and carbonate sediments and their eventual uplift and weathering. Because the weathering of organic C sediments involves the oxidation of reduced C to carbon dioxide, it is represented in Figure 6.7 as a source of C to the atmosphere. The weathering of carbonate sediments, on the other hand, leads to the dissolution of carbonates into rivers and streams and their eventual transfer to the ocean, and thus this process is represented as a source of C to the surface ocean. It should be noted, however, that while the carbonate–sediment formation and weathering steps in Figure 6.7 show no coupling to the atmosphere, these steps are in fact coupled to the atmosphere. Recall from Chapter 2 that because the ocean pH is about 8, the major form of dissolved inorganic C in the oceans is bicarbonate ion (HCO_3^-). For this reason, the precipitation of calcium carbonate causes the production of gaseous carbon dioxide and thus the transfer of C from the atmosphere to the ocean; that is,

$$Ca^{2+} + 2HCO_3^- \rightarrow (CaCO_3)_s + H_2O + (CO_2)_g \tag{R6.1}$$

Conversely, the dissolution of calcium carbonate requires the simultaneous dissolution of gaseous carbon dioxide and thus the transfer of C from the atmosphere to the ocean; that is,

$$(CaCO_3)_s + H_2O + (CO_2)_g \rightarrow Ca^{2+} + 2HCO_3^- \tag{R6.2}$$

It follows therefore that, in addition to being a sink of deep ocean C, the formation of carbonate sediments transfers 0.2 Gton C year^{-1} from the surface ocean to the atmosphere, and the weathering of these same sediments takes 0.2 Gton C year^{-1} from the atmosphere and transfers it to the surface ocean.

---

[3]Note that because our focus here is on C, we can use the simple reactions for photosynthesis and respiration involving "CH$_2$O."

Finally note that our representation of the preindustrial C cycle in Figure 6.7 is in steady state, with the total source of C for each reservoir exactly balanced by the sum of all its sinks. For example, we have assumed that the gross primary production rate of the terrestrial biosphere of 96 Gtons C year^{-1} is balanced by the sum of respiration and literfall/decay. In the ocean, on the other hand, the gross ocean surface production rate of 40 Gtons C year^{-1} is balanced by remineralization of organic C in the surface ocean at a rate of 36 Gtons C year^{-1}, and sinking of organic C out of the surface ocean at a rate of 4 Gtons C year^{-1}. Of this 4 Gtons C year^{-1}, 3.9 Gton C are remineralized annually in the deep ocean and eventually returned to the surface ocean by upwelling and 0.1 Gtons C year^{-1} falls to the ocean bottom and form organic C sediments.

On the basis of the data in Figure 6.7, the **K** matrix for the steady-state, preindustrial C cycle is given by

$$
\mathbf{K} = \begin{bmatrix}
-0.2933 & 0.0578 & 0.032 & 0.0799 & 0.0 & 0.0 & 5.0 \times 10^{-9} & 0.0 \\
0.16 & -0.1156 & 0.0 & 0.0 & 0.0 & 0.0 & 0.0 & 0.0 \\
0.0 & 0.0578 & -0.032 & 0.0 & 0.0 & 0.0 & 0.0 & 0.0 \\
0.1333 & 0.0 & 0.0 & -0.1532 & 9.737 \times 10^{-4} & 20.0 & 0.0 & 2.85 \times 10^{-9} \\
0.0 & 0.0 & 0.0 & 0.0333 & -9.79 \times 10^{-4} & 2.167 & 0.0 & 0.0 \\
0.0 & 0.0 & 0.0 & 0.04 & 0.0 & 0.0 & 0.0 & 0.0 \\
0.0 & 0.0 & 0.0 & 0.0 & 0.0 & 0.0556 & -5.0 \times 10^{-9} & 0.0 \\
0.0 & 0.0 & 0.0 & 0.0 & 5.263 \times 10^{-6} & 0.0 & 0.0 & -2.85 \times 10^{-9}
\end{bmatrix}
$$

In the next sections, we will use this model as the starting point for investigating the response of the global C cycle to the sizable inputs of atmospheric CO_2 that have occurred over the past century.

6.4. THE IMPACT OF ANTHROPOGENIC EMISSIONS AND THE "AIRBORNE FRACTION"

Since the preindustrial era, atmospheric CO_2 concentrations have increased from about 280 ppmv to about 355 ppmv. These concentrations correspond to atmospheric C contents of 600 and 755 Gtons, respectively. And thus the atmospheric reservoir of C has increased in size by about 155 Gtons. On the other hand, if we integrate the curves for anthropogenic CO_2 emissions illustrated in Figure 6.1 from 1860 to 1990, we find that the total amount of excess C placed in the atmosphere by human activities is about 350 Gtons (i.e., 230 Gtons C from fossil fuel burning and 120 Gtons C from deforestation of mostly tropical forests). The ratio of these two numbers defines the fraction of the anthropogenic CO_2 emissions that have remained in the atmosphere. This ratio is referred to as the *airborne fraction* of atmospheric CO_2 and is approximately 40%; that is,

$$\text{Airborne fraction} \approx \frac{155 \text{ Gtons}}{350 \text{ Gtons}} \approx 0.44 \tag{6.4}$$

Thus we find that for each Gton of CO_2 added to the atmosphere by anthropogenic activities, only about 0.4 Gton remains in the atmosphere while 0.6 Gton is removed and placed somewhere else in the earth system. What are the processes responsible for removing 60% of the excess CO_2 emissions and where does this C end up residing? Let's use *BOXES* and our eight-box model of the C cycle to try to answer these questions.

6.4.1. Simple *BOXES* Simulation

To simulate the effect of anthropogenic CO_2 emissions in the global C cycle, we first need to decide how we will represent these emissions. As it turns out, the actual anthropogenic CO_2 emissions from fossil fuel burning and deforestation (illustrated in Figure 6.8A) describe a complicated function of time that is not easily simulated in *BOXES*, with its constant k-values. We can simplify matters somewhat by making two approximations. We assume that all anthropogenic emissions arise from a single source, namely, the transfer of C from the organic-C sediment (i.e., fossil fuel burning), and we approximate these emissions with a series of "step functions." For the first 60 years of the simulation, nominally representing the period from 1860 to 1920, we assume a constant rate of anthropogenic CO_2 emissions of 1 Gton C year^{-1}. For the next 40 years from 1920 to 1960, we assume a constant emission rate of 3.5 Gtons C yr^{-1}, and for the last 30 years from 1960 to 1990, we assume a constant emission rate of 5 Gtons C year^{-1}. Note in Figure 6.8A that although these functions do not exactly reproduce the CO_2 emission trends over the past 130 years, they represent a reasonable facsimile of them. Moreover, it is easy to show that the cumulative emissions from these step functions over the 130-year period is 350 Gtons C year^{-1}, the actual amount estimated to have come from anthropogenic emissions since 1860.

The adoption of this step function representation for the emissions allows us to simulate the effect of anthropogenic activities using *BOXES* in three consecutive stages, each requiring the adjustment of only one nondiagonal element of the **K** matrix (see Table 6.1). In "Stage 1," we simulate the period from 1860 to 1920. Since the calculation begins in 1860, the initial reservoir contents for this simulation are just the values from the preindustrial, steady-state cycle. To account for the anthropogenic CO_2 emissions during this period, we demand that $F_{7\to1}$, the flux from the organic sediments to the atmosphere, be given by

$$F_{7\to1} = F_{\text{Weathering}} + F_{\text{Anthropogenic}}$$

$$= (0.1 + 1) \text{ Gton C year}^{-1}$$

$$= 1.1 \text{ Gton C year}^{-1} \tag{6.5}$$

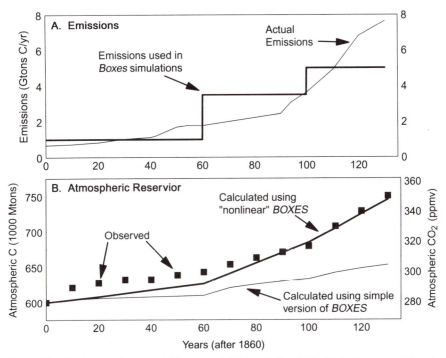

Figure 6.8. The impact of human activities on the global C cycle since 1860. (A) Anthropogenic CO_2 emissions since 1860, with the thin solid line indicating the actual, total emissions estimated from fossil fuel burning and deforestation (see Figure 6.1) and the heavy solid line indicating the anthropogenic emissions adopted in the *BOXES* simulations. (B) The atmospheric C reservoir (left-hand axis) and CO_2 mixing ratio (right-hand axis) as a function of year since 1860, with the boxes indicating the observed values (see Figure 6.1), the thin solid line indicating the results using the simple *BOXES* application described in Section 6.4.1, and the heavy solid line indicating the results using the pseudo-nonlinear application of *BOXES* described in Section 6.4.2.

and this requires that we change $k_{7\rightarrow1}$ from its preindustrial value of 5×10^{-9} to 5.5×10^{-8} year^{-1}. Thus the new **K** matrix for Stage 1 of the simulation is given by

$$\mathbf{K} = \begin{bmatrix}
-0.2933 & 0.0578 & 0.032 & 0.0799 & 0.0 & 0.0 & 5.5 \times 10^{-8} & 0.0 \\
0.16 & -0.1156 & 0.0 & 0.0 & 0.0 & 0.0 & 0.0 & 0.0 \\
0.0 & 0.0578 & -0.032 & 0.0 & 0.0 & 0.0 & 0.0 & 0.0 \\
0.1333 & 0.0 & 0.0 & -0.1532 & 9.737 \times 10^{-4} & 20.0 & 0.0 & 2.85 \times 10^{-9} \\
0.0 & 0.0 & 0.0 & 0.0333 & -9.79 \times 10^{-4} & 2.167 & 0.0 & 0.0 \\
0.0 & 0.0 & 0.0 & 0.04 & 0.0 & -22.22 & 0.0 & 0.0 \\
0.0 & 0.0 & 0.0 & 0.0 & 0.0 & 0.0556 & -5.5 \times 10^{-8} & 0.0 \\
0.0 & 0.0 & 0.0 & 0.0 & 5.263 \times 10^{-6} & 0.0 & 0.0 & -2.85 \times 10^{-9}
\end{bmatrix}$$

TABLE 6.1.
Fluxes and k Values Used in BOXES Simulations of the Effects of Anthropogenic CO_2 Emissions on the Global C Cycle

Time Period	Anthropogenic CO_2 Emissions (Gtons year^{-1})	Simple BOXES Version $k_{7\to1}$ (year^{-1})
1860–1920 (Stage 1)	1	5.5×10^{-8}
1920–1960 (Stage 2)	3.5	1.8×10^{-7}
1960–1990 (Stage 3)	5	2.55×10^{-7}

	Pseudo-nonlinear BOXES Version		
Time Period	$k_{7\to1}$ (year^{-1})	$k_{1\to2}$ (year^{-1})	$k_{1\to4}$ (year^{-1})
1860–1870	5.5×10^{-8}	0.16	0.1333
1870–1880	5.5×10^{-8}	0.159	0.1325
1880–1890	5.5×10^{-8}	0.158	0.1317
1890–1900	5.5×10^{-8}	0.157	0.1307
1900–1910	5.5×10^{-8}	0.156	0.1298
1910–1920	5.5×10^{-8}	0.155	0.1289
1920–1930	1.8×10^{-7}	0.1546	0.1281
1930–1940	1.8×10^{-7}	0.1522	0.1255
1940–1950	1.8×10^{-7}	0.1498	0.1230
1950–1960	1.8×10^{-7}	0.1475	0.1206
1960–1970	2.55×10^{-7}	0.1454	0.1183
1970–1980	2.55×10^{-7}	0.1425	0.1153
1980–1990	2.55×10^{-7}	0.1397	0.1124

where the bold numbers indicate those elements of the **K** matrix that have been changed from the steady-state case.

"Stage 2" of the simulation takes us from 1920 to 1960. We now use the final reservoir contents from the Stage 1 calculation (i.e., $t = 60$ years) to specify the initial reservoir contents; and to specify an anthropogenic CO_2 emission rate of 3.5 Gtons C year^{-1}, we now set $k_{7\rightarrow1}$ at 1.8×10^{-7} year^{-1}. Finally for "Stage 3" from 1960 to 1990, we use the Stage 2 final reservoir contents to specify our initial conditions and we set $k_{7\rightarrow1}$ at 2.55×10^{-7} year^{-1} to get an anthropogenic emission rate of 5 Gtons C year^{-1}.

The resulting atmospheric C reservoir as a function of time is illustrated in Figure 6.8B. Comparison of this result with measurements of the concentration of atmospheric CO_2 during this time period reveals a significant discrepancy. While the actual atmospheric C reservoir increased from roughly 600 to about 755 Gtons C over the 130 time period, our model predicts an increase of about one-third this amount, from 600 to only about 650 Gtons C. Clearly our model has a problem. We will have to identify and correct this problem, if we hope to use *BOXES* to analyze the earth system's response to anthropogenic CO_2 emissions. We will attempt to do this in the next section.

6.4.2. A "Pseudo-nonlinear" Simulation Using *BOXES*

Among the numerous simplifications in our treatment of the global C cycle in the previous section, probably the most serious is the assumption of linearity between the abundance of C in the atmospheric reservoir and the flux of C to the ocean and the terrestrial biosphere. Since $k_{1\rightarrow2}$ and $k_{1\rightarrow4}$, which control the fluxes from the atmosphere to the terrestrial biosphere and the ocean, respectively, were not changed throughout the simulation, any change in the size of the atmospheric reservoir leads to a corresponding change in the fluxes from the atmosphere to these other two reservoirs. In fact, this turns out to be a rather poor representation of how the C cycle actually operates. Let's consider the exchange between the atmosphere and ocean first.

6.4.2.1. Sensitivity of Oceanic C Reservoir—Derivation of the Revelle Factor

Imagine that $p(CO_2)$, the partial pressure of atmospheric CO_2, is increased by some amount $\Delta p(CO_2)$ and this causes C_C, the total concentration of dissolved carbon in the ocean, to increase by an amount given by ΔC_C. We define ϵ as the ratio of the relative change in $p(CO_2)$ to the relative change in C_C:

$$\epsilon = \frac{\dfrac{\Delta p(CO_2)}{p(CO_2)}}{\dfrac{\Delta C_C}{C_C}} \tag{6.6}$$

This ratio is sometimes referred to as the *Revelle factor*; it is named after the geochemist Roger Revelle, who did much of the seminal work on the exchange of CO_2 between the atmosphere and ocean in the 1950s, 1960s, and 1970s.

To better understand the Revelle factor and its implications for the relationship between changes in $p(CO_2)$ and C_C, let's first review some material from Chapter 2 concerning the aqueous carbonate system. In Section 2.5.1 of that chapter we derived an expression for the relationship between $p(CO_2)$ and C_C in aqueous solution as a function of pH:

$$C_C = K_{H,CO_2} p(CO_2) \frac{[H^+]^2 + K_{H2CO3,1}[H^+] + K_{H2CO3,2}K_{2,H2CO3}}{[H^+]^2} \qquad (2.5.4)$$

At oceanic pH values it is easily shown that Equation (2.5.4) can be approximated by

$$C_C \approx p(CO_2) \frac{K_{H,CO_2}K_{H2CO3,1}}{[H^+]} \qquad (6.7)$$

Thus we see that, at constant pH, C_C is proportional to $p(CO_2)$ and thus we might at first guess that ϵ must be equal to unity; that is, a doubling in $p(CO_2)$ would cause a doubling in C_C. The fallacy in this argument arises from the assumption of constant pH. Because $[H^+]$ increases as CO_2 dissolves in water and forms carbonic acid and its conjugate bases, there cannot be a 1:1 linear relationship between C_C and $p(CO_2)$. More specifically, since $[H^+]$ increases with the dissolution of CO_2, C_C must increase less rapidly than $p(CO_2)$. Accordingly the Revelle factor ϵ, as defined by Equation (6.6), must be greater than unity.

To determine a specific value for ϵ we must consider the following processes:

1. The equilibria of the CO_2 carbonate system

$$(CO_2)_g \rightleftarrows H_2CO_3^* \qquad (R2.7')$$

$$H_2CO_3^* \rightleftarrows HCO_3^- + H^+ \qquad (R2.9')$$

$$HCO_3^- \rightleftarrows CO_3^{2-} + H^- \qquad (R2.10)$$

where, recall from Chapter 2, $H_2CO_3^* = (CO_2)_{aq} + H_2CO_3$ is the total amount of undissociated carbon dioxide in solution.

2. The equilibrium from the acid dissociation of H_2O

$$H_2O \rightleftarrows H^+ + OH^- \qquad (R2.2)$$

3. The equilibrium from the acid dissociation of boric acid

$$B(OH)_3 + H_2O \rightleftarrows H^+ + B(OH)_4^- \qquad (R6.3)$$

which, at a total boric acid concentration in seawater of 4.1×10^{-4} M, has a small but non-negligible buffering effect on pH.

4. The equation describing charge balance for seawater

$$[H^+] + [Alk] = [HCO_3^-] + 2[CO_3^{2-}] + [B(OH)_4^-] + [OH^-] \qquad (6.8)$$

where $[Alk] = 2.5 \times 10^{-3}$ Eq liter^{-1} is the alkalinity of seawater—that is, the concentration of excess strong base cations minus strong acid anions.

Using the formulation developed in Chapter 2, we can derive a set of six coupled algebraic equations, and, from these equations, we can solve for $C_C = [H_2CO_3^*] + [HCO_3^-] + [CO_3^{2-}]$ as a function of $p(CO_2)$. The results, which are illustrated in Figure 6.9, can then be used to derive the Revelle factor. A value of about 10 is thus obtained:

$$\epsilon = \frac{\dfrac{\Delta p(CO_2)}{p(CO_2)}}{\dfrac{\Delta C_C}{C_C}} \approx \frac{\dfrac{355\ \text{ppmv} - 280\ \text{ppmv}}{280\ \text{ppmv}}}{\dfrac{C_C(355\ \text{ppmv}) - C_C(280\ \text{ppmv})}{C_C(280\ \text{ppmv})}}$$

$$= \frac{0.27}{0.021} \approx 10 \qquad (6.9)$$

A Revelle factor of 10 means that for each 1% change in atmospheric CO_2, oceanic C changes by only 0.1%. This is a very different dependence from that used in our previous *BOXES* simulation, where a 1% change in the atmospheric reservoir, C_1, causes a 1% change in the flux of carbon from the atmosphere to the ocean and ultimately, therefore, a 1% increase in the size of the ocean reservoirs. If we hope to simulate the global C cycle with any degree of accuracy, we must account for this effect. However, doing this presents a problem since the numerical method in *BOXES* implicitly assumes a one-for-one correspondence between a change in a reservoir amount and the flux out of that reservoir.

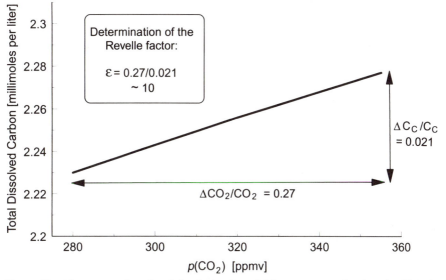

Figure 6.9. The concentration of total dissolved carbon (C_C) in seawater as a function of $p(CO_2)$, the partial pressure of atmospheric carbon dioxide. The ratio of the relative change in $p(CO_2)$ to the relative change in C_C from these results implies a Revelle factor, ϵ, of about 10. (Note that the results illustrated here were obtained using equilibrium constants appropriate for seawater that account for ionic strength effects; that is, $K_{H,CO_2} = 0.048$ M atm^{-1}, $K_{H_2CO_3,1} = 8.8 \times 10^{-7}$ M, $K_{H_2CO_3,2} = 5.6 \times 10^{-10}$ M, and $K_{H_3BO_3,1} = 1.6 \times 10^{-9}$.)

Fortunately it is not an insurmountable problem. We of course have the option of using more sophisticated numerical methods that explicitly allow for nonlinear terms.[4] We also have the option of using *BOXES*, but it will require some extra work on our part. To account for the Revelle factor using *BOXES*, we must periodically stop the calculation, reduce the value for $k_{1\to4}$ as C_1 increases, so that the relative change in the flux of C from the atmosphere to the ocean is only one-tenth of the relative change in C_1, and restart the calculation again. For example, suppose we run *BOXES* from some time t to a later time $t + \Delta t$ and, in that interval, C_1 changes from a value of $C_1(t)$ to a value of $C_1(t + \Delta t)$. Further suppose that $k_{1\to4}°$ is the k value used for the calculation over this time period. In order to limit the increase in the flux to one-tenth of the change in C_1, it is easy to show that $k_{1\to4}'$, the k value for the next portion of the calculation (i.e., from $t + \Delta t$ to $t + 2\Delta t$), should be adjusted to

$$k'_{1\to4} = k°_{1\to4}\left(0.1 + 0.9\,\frac{C_1(t)}{C_1(t + \Delta t)}\right) \tag{6.10}$$

where the parameters of 0.9 and 0.1 come directly from the value we have derived for ϵ of 10. As in most numerical integration schemes, the value of Δt needed to give satisfactory results depends upon the characteristic time scales of the system being simulated and also upon the desired accuracy of the simulation. After some experimentation, we have found that a Δt of 10 years gives reasonably accurate results for this problem.

6.4.2.2. Sensitivity of the Terrestrial Biosphere—The Beta Factor

Just as the Revelle factor is used by marine geochemists to describe the relationship between atmospheric CO_2 and ocean C, terrestrial ecologists refer to the *beta factor* to denote the relationship between atmospheric CO_2 and C storage in the terrestrial biosphere. However, in contrast to the Revelle factor, the beta factor is generally defined with the relative change in atmospheric CO_2 in the denominator rather than the numerator:

$$\beta = \frac{\dfrac{\Delta C_T}{C_T}}{\dfrac{\Delta p(CO_2)}{p(CO_2)}} \tag{6.11}$$

where C_T is the total amount of C stored in the living and dead terrestrial biospheres.

Defining a value for β is a daunting task that requires integrating complex biological processes acting over scales ranging from the cellular to the global. Not surprisingly therefore, its value is considerably more uncertain than that of the Revelle factor. Simple exposure experiments, where green plants are exposed to enhanced concentrations of CO_2 under controlled conditions, tend to confirm the existence of a

---

[4]The ambitious student can pursue this option by developing his or her own code to integrate the differential equations. J. C. G. Walker's *Numerical Methods in the Earth Sciences* (Cambridge University Press, 1993) provides a number of useful routines. Less ambitious students who use Macintosh systems have the option of using STELLA, a software package designed to solve any arbitrary system of coupled, differential equations.

CO₂ fertilization effect, with a doubling in ambient CO_2 concentration generally causing a 20–40% increase in photosynthetic rates. However, these results do not translate directly into a value for β, since the amount of C that ultimately gets stored in the terrestrial biosphere is a function of the rates of respiration and decay, as well as photosynthesis. In fact, because increasing CO_2 concentrations tend to increase surface temperatures and the rate of decay of soil organic matter tends to increase with increasing temperatures, a negative value for β is not out of the realm of possibility.

Recognizing the uncertainties inherent in our approach, let's assume that the exposure experiments can be used to define β and adopt a central value of 0.3 for this parameter. Then, following the same approach as that described for $k_{1\to4}$, we can derive an equation for the adjustment required in $k_{1\to2}$ for each Δt:

$$k'_{1\to2} = k°_{1\to2} \left(0.3 + 0.7 \, \frac{C_1(t)}{C_1(t + \Delta t)} \right) \tag{6.12}$$

6.4.2.3. Results Using the "Pseudo-nonlinear" Model

In the simple version of *BOXES* described in Section 6.3, we simulated the 130-year period from 1860 to 1990 in three separate stages so that we could vary the rate of anthropogenic CO_2 emissions in a manner that approximated the actual emissions. For the pseudo-nonlinear version of *BOXES*, we need to carry out the simulation in 13 separate stages, each lasting 10 years, so that we can account for the nonlinear aspects of the exchange of C from the atmosphere. As in the simple version, the initial reservoir contents for each of the 13 stages are determined by the final amounts obtained from the previous stage, and the values for $k_{7\to1}$ are specified to give the appropriate input of anthropogenic CO_2. In addition, the values for $k_{1\to4}$ and $k_{1\to2}$ are adjusted for each stage using Equations (6.10) and (6.12), respectively, along with the solution for C_1 from the previous stage. The k values used for each of the 13 stages are listed in Table 6.1.

The heavy solid line in Figure 6.8B illustrates the resulting solution for the atmospheric reservoir as a function of time. In stark contrast to the simple version of *BOXES*, we now find excellent agreement between our calculations and the actual measured variations in atmospheric CO_2. For example, in 1990 the pseudo-nonlinear version of *BOXES* predicted an atmospheric reservoir of 745 Gtons C while the observed CO_2 mixing ratio in the early 1990s of 355 ppmv yielded a reservoir size of 755 Gtons C. By comparison, the simple *BOXES* version predicted a 1990 atmospheric reservoir of only 650 Gtons C. Clearly the inclusion of the nonlinear aspects of the exchange of C from the atmosphere to the ocean and the terrestrial biosphere greatly improves our simulation of the global C cycle. More importantly, it allows us to reasonably reproduce the magnitude of the CO_2 increase over the past 100 years or so, and thus we are able to correctly predict an airborne fraction for anthropogenic CO_2 of 40%. In the next section we will use our pseudo-nonlinear version of *BOXES* to address the other question we had posed at the outset of this section: Where has the 60% nonairborne fraction gone?

6.4.2.4. The Fate on the Nonairborne Fraction and the "Missing Sink"

Table 6.2 lists the contents of the eight-reservoirs of our C cycle at the beginning (i.e., 1860) and the end (i.e., 1990) of the 130-year simulation. Our model predicts that about 82 Gtons C—or 23% of the C initially placed into the atmosphere by anthropogenic emissions since 1860—went into the ocean, with about two-thirds of that amount in the deep ocean and the rest in the surface ocean. A somewhat larger portion, 128 Gton C—or 35% of the anthropogenic emissions—ended up in the terrestrial biosphere according to our model, with approximately equal amounts in the living and dead portions. Although not exactly the same, the results from more sophisticated models of the global C cycle tend to be reasonably consistent with those we have obtained from our pseudo-nonlinear version of *BOXES* (see Table 6.3).

The bottom line of these analyses is that, in order to balance the global C budget in the face of the sizable emissions of anthropogenic CO_2 over the past 100 years or so, both the ocean C reservoir and the terrestrial biosphere must have grown over this period by about 100 Gtons C. The requirement for an increase of this magnitude in the oceans does not appear to present a problem for marine geochemists, who estimate that an excess of about 2 Gtons C presently goes into the ocean each year. On the other hand, the requirement for a 100-Gton increase in the terrestrial biosphere presents a significant problem for terrestrial ecologists. In the first place, large portions of the terrestrial biosphere have been destroyed over the past 100 years, because a rapidly expanding population of humans has sought out new lands for agriculture and urban/industrial development. In fact, it is estimated that the clearing and burning of forests since the mid-1800s have actually added (not removed) about 120 Gtons C to the atmosphere. Most of this deforestation has occurred in the tropics and in the latter half of the twentieth century. Thus balancing the C budget requires extratropical forests to have grown by about 200 Gtons C

TABLE 6.2
The Fate of the Anthropogenic CO_2 Emissions According to the Pseudo-nonlinear Version of *BOXES*

Reservoir	Amount Assumed in 1860[a] (Gtons C)	Amount Calculated in 1990[b] (Gtons C)	Change (Gtons C)	Percentage of 350 Gtons C Emitted
Atmosphere	600	745[c]	145	41
Living terrestrial biosphere	830	890	60	17
Dead terrestrial biosphere	1500	1568	68	19
Surface ocean	1000	1026	26	7
Living marine biosphere	1.8	1.85	0.05	—
Deep ocean	38,000	38,056	56	16
Organic C sediment	20,000,000	20,000,000	—	—
Carbonate sediment	70,000,000	70,000,000	—	—

[a]Based on preindustrial, steady-state model.
[b]Obtained from pseudo-nonlinear version of *BOXES*.
[c]Actual value is 755 Gtons C.

TABLE 6.3
The Budget and Fate of Anthropogenic CO_2 Emissions During the 1980s Based on Best Available Science

	Gtons C year^{-1}	Percentage of Total Source
A. Anthropogenic sources		
Emissions from fossil fuel burning and cement manufacture	5.5	77
Emissions from clearing of tropical forests	1.6	23
Total anthropogenic emissions	7.1	100
B. Partitioning among reservoirs		
Atmosphere	3.2	44[a]
Ocean	2.0	28[a]
Terrestrial biosphere		
Regrowth of forests in Northern Hemisphere	0.5	0.07[a]
Additional terrestrial sink required to balance budget[b]	1.4	20[a]

[a]Note similarity of these percentages with those obtained from our pseudo-nonlinear version of *BOXES* listed in Table 6.2.
[b]This is the so-called "missing sink."

Source: Intergovernmental Panel on Climate Change, *Climate Change 1994: Radiative Forcing of Climate Change,* Cambridge University Press, New York, 1995.

(i.e., 100 Gtons C from the nonairborne fraction that did not go into the ocean and an additional 100 Gtons C to counteract the effect of deforestation). Although there is evidence for an expansion/regrowth of temperate and boreal forests in the Northern Hemisphere since the turn of the century (see Table 6.3), the amount of excess C typically stored in these forest appears to be at most 100 Gtons. This leaves about 100 Gtons of C unaccounted for—the so-called *missing sink* of C. Where is the missing sink? Have ocean models underestimated the rate of CO_2 exchange into the ocean and is the missing sink therefore in the ocean? Or have terrestrial ecologists underestimated the amount of C going into temperate and boreal forests? These questions represent a major puzzle for scientists studying the global C cycle and will likely remain so for some time. Unfortunately, until these questions can be satisfactorily answered, we must also question our ability to predict how the global C cycle will respond to continued anthropogenic CO_2 emissions in the future (see Text Box 6.2).

6.5. THE PERSISTENCE OF THE ANTHROPOGENIC PERTURBATION

Suppose, perhaps because of fears of global warming, humankind were to decide to immediately stop all activities that caused CO_2 to be emitted into the atmosphere. How long would it take for the excess CO_2 that had accumulated in the atmosphere since the Industrial Revolution to dissipate? We can address this ques-

TEXT BOX 6.2	THE AIRBORNE FRACTION: WILL IT CHANGE IN THE FUTURE?

Historically the airborne fraction of anthropogenic CO_2 has been about 0.4. If this fraction never changes, then predicting how atmospheric CO_2 concentrations will respond to continued anthropogenic emissions becomes a trivial problem. However, it is not at all clear that this will be the case. Consider, for example, the response of the terrestrial biosphere. As discussed previously, there is very little evidence to support the beta factor value of 0.3 we assumed in our model calculation. But, even if we accept this value as being appropriate for the past century, there is no reason to assume that this value will continue to apply in the coming decades. Insufficient supply of other nutrients may eventually limit photosynthetic rates, preventing further enhancements in photosynthesis from the CO_2 fertilization effect. Alternatively, increasing temperatures caused by an enhanced greenhouse effect may accelerate the rate of decay of dead organic matter in soils, overwhelming the CO_2 fertilization effect and causing the excess C that had accumulated in the terrestrial biosphere to be suddenly transferred back to the atmosphere. Either of these scenarios would cause atmospheric CO_2 concentrations to build up at rates significantly faster than would be predicted on the basis of an airborne fraction of 0.4 and would therefore have the potential to greatly exacerbate any climatic perturbation that might ensue. Evaluation of the sensitivity of future CO_2 concentrations to variation in the beta factor has been left as an exercise in the problem section at the end of the chapter.

tion using *BOXES*. We simply have to reset $k_{7 \rightarrow 1}$ back to its preindustrial value of 5×10^{-9} year^{-1}, specify initial reservoir amounts equal to the values we calculated at the end of the 130-year simulation from the previous section, and run the model. [For the pseudo-nonlinear version of *BOXES*, this also involves running the model in 10-year stages and adjusting the values of $k_{1 \rightarrow 4}$ and $k_{1 \rightarrow 2}$ according to Equations (6.10) and (6.12) as described in Section 6.4.]

The resulting atmospheric C reservoir calculated as a function of time for the simple and pseudo-nonlinear versions of *BOXES* are illustrated in Figure 6.10. While the absolute amount of C that leaves the atmosphere for the nonlinear version of *BOXES* is clearly much larger than that for the simple version, the *e*-folding time for the decay of the excess atmospheric C is pretty much the same for both models—of the order of several decades. This result is somewhat counterintuitive. A standard calculation of the residence time for atmospheric C (obtained by dividing the atmospheric reservoir by the sum of the fluxes leaving the atmosphere) yields a value of less than 10 years, and yet the time required to flush the atmosphere of anthropogenic C is several decades. This substantially longer time arises from the fact that the anthropogenic CO_2 that had been added to the system has already been partitioned between the atmosphere, the ocean, and the terrestrial biosphere. As a result, the flow of CO_2 out of the atmosphere is largely controlled by the longer time scales involved with mixing into the deep ocean and exchange with the dead terrestrial biosphere.

The relatively long time scale required to remove the excess anthropogenic CO_2

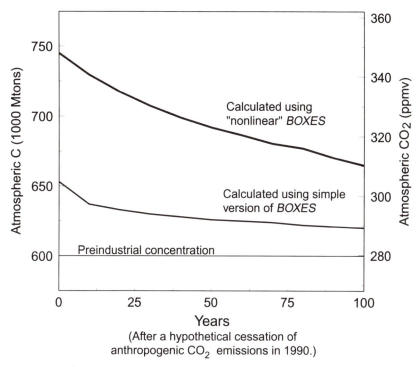

Figure 6.10. The calculated decay in the reservoir of atmospheric C after a hypothetical cessation of anthropogenic CO_2 emissions in 1990. Results are shown for the simple *BOXES* version (thin solid line) and the pseudo-nonlinear *BOXES* version (heavy solid line).

from the atmosphere adds some urgency to the debate over global warming. Should it be discovered that the increase in CO_2 and other greenhouse gases is causing significant and undesirable perturbations to the climate, it will take decades to reverse the effect, even if all CO_2 emissions were instantaneously halted.

6.6. CONCLUSION

For the biogeochemist, the study of the global C cycle is a humbling experience. While the world debates over what to do about rising concentrations of atmospheric CO_2, we remain unable to provide definitive answers to key scientific questions that lie at the core of the debate: (i) Where does the C emitted into the atmosphere over the past century currently reside? (ii) How will the biosphere respond to continued increases in atmospheric CO_2? and (iii) What will be the trajectory of future CO_2 concentrations if anthropogenic emissions continue?

The study of the C cycle also provides a cautionary lesson for the students of biogeochemistry. Although it is usually possible to construct a simple linear model of a global biogeochemical cycle, there is no guarantee that the model will capture important, and in some cases essential, features of the cycle that ultimately determine how the system will respond to perturbations. For this reason, we must always proceed with great care in using the results from model calculations like those from *BOXES*, and in some cases we must take extra steps to ensure that the results will be a reasonable reflection of the actual earth system.

SUGGESTED READING

Bolin, B., The carbon cycle, *Scientific American*, **223**, 124–132, 1970.

Intergovernmental Panel on Climate Change, *Climate Change 1994: Radiative Forcing of Climate Change*, Cambridge University Press, New York, 1995.

Moore, B., III, and B. H. Braswell, The lifetime of excess atmospheric carbon dioxide, *Global Biogeochemical Cycle*, **8**, 23–38, 1994.

Siegenthaler, U., and J. L. Sarmiento, Atmospheric carbon dioxide and the ocean, *Nature*, **365**, 119–125, 1993.

Stryer, L., *Biochemistry*, second edition, W. H. Freeman, New York, 1981.

Woodward, F. I., Predicting plant responses to global environmental change, *Plant Physiologist*, **122**, 239–251, 1992.

PROBLEMS

1. The Framework Convention on Climate Change of the United Nations adopts the long-term objective of stabilizing atmospheric concentrations of all greenhouse gases at levels ". . . that would prevent dangerous anthropogenic interference with the climate system." Suppose it were determined that the prevention of "dangerous interference with the climate system" required that CO_2 concentrations be kept below 400 ppbv. Use *BOXES* to determine the maximum anthropogenic emission rate allowed under the Convention.

2. How would the results obtained in Problem 1 change if the beta factor were to change to 0.15? to 0?

The Global Sulfur Cycle

7

7.1. INTRODUCTION

Sulfur (S) is a nonmetallic element in Group VIA of the Periodic Table. As summarized in Figure 7.1, it has an atomic number of 16, an atomic weight of 32.06,
and six electrons in its outer shell. Like phosphorus (P), S is found in living organisms in small amounts (i.e., about 0.25%) and plays a crucial role in the biosyn-
thetic processes of most organisms. However, the similarity ends there. Unlike P,
dissolved S is in plentiful supply in most natural waters; as a result, S is rarely, if
ever, a limiting nutrient. Moreover, while P is only found in one oxidation state (+5)
within natural environments, S is found in a number of different oxidation states.
And, as we shall see, it is the facility with which S can be transformed from one
oxidation state to another that gives the element its important role in the biogeochemistry of the earth system.

7.2. S REDOX CHEMISTRY

Because S has six electrons in its outer shell, it can occupy any one of eight oxidation states ranging from its most reduced state of -2 to its most oxidized of $+6$.
As it turns out, S compounds within the earth system are commonly found in six
of these oxidation states. Some of the more prevalent of these S compounds are
listed in Table 7.1. A noteworthy feature of the list in Table 7.1 is the fact that al-

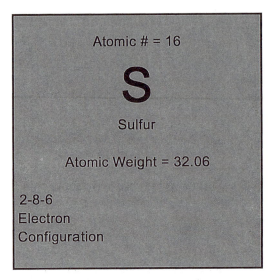

Figure 7.1. Sulfur is a nonmetallic element in Group VIA of the Periodic Table.

though S is found in a variety of oxidation states in the atmosphere, hydrosphere, and lithosphere, it is predominantly in the -2 oxidation state within organic compounds.

The thermodynamic relationships between the various oxidation states of S are summarized in Figure 7.2. From these data and those of the Appendix, it is possible to derive a series of equations describing the stability boundaries for the various oxidation-state couplets of S. Lines for seven of these couplets are drawn on a pe–pH diagram in Figure 7.3A. The figure portrays a very complicated situation with lines crisscrossing repeatedly. Fortunately, however, the vast majority of these

TABLE 7.1
Naturally Occurring S Compounds and Their Oxidation States[a]

	Atmosphere		Hydrosphere	Lithosphere		Biosphere
	Gas	Particulate		Soils	Rocks	
-2	H_2S, RSH, RSR, OCS, CS_2	—	H_2S, HS^-, S^{2-}, RS^-	S^{2-}, HS^- MS	S^{2-}, HgS	Methionine, cysteine
-1	RSSR	—	RSSR	SS^{2-}	FeS_2	—
0	CH_3SOCH_3	—	—	S_8	—	—
$+2$	—	—	$S_2O_3^{2-}$	—	—	—
$+4$	SO_2	$SO_2 \cdot H_2O$, CH_3SOOH	H_2SO_3, HSO_3^-, SO_3^{2-}	SO_3^{2-}	—	—
$+6$	SO_3	H_2SO_4, NH_4HSO_4, $(NH_4)_2SO_4$, Na_2SO_4, CH_3SO_3H	H_2SO_4, HSO_4^-, SO_4^{2-}	$CaSO_4$	$CaSO_4 \cdot 2H_2O$, $MgSO_4$	—

[a]"R" denotes an organic radical (e.g., CH_3), and "M" denotes a metal ion.

Source: Charlson, R. J., T. L. Anderson, and R. E. McDuff, The sulfur cycle, in *Global Biogeochemical Cycles*, edited by S. Butcher, R. J. Charlson, G. Orians, and G. V. Wolfe, Academic Press, New York, 1992.

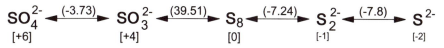

Figure 7.2. The oxidation states of S. The number in square brackets beneath each sulfur-bearing compound indicates the oxidation state of S within the compound, and the number in parentheses between each pair of compounds indicates the log of the equilibrium constant for the redox half-reaction between the two compounds. (Note that S in the 0 oxidation state is most often found as S_8, a crystal with a rhombic structure.)

lines describe metastable boundaries. When we omit these metastable boundaries and only consider the region of pe–pH space where water is stable, we obtain Figure 7.3B, a vastly simpler picture with essentially only three stability boundaries involving S in the +6, 0, and −2 oxidation states.

The results illustrated in Figure 7.3B indicate that for the major portion of the stability region of water, there are only two stable oxidation states of S: the −2 oxidation state in the bottom portion of the stability region and the +6 oxidation state in the upper portion. (The neutral oxidation state of S is also stable, but only under highly acidic conditions and over a relatively small portion of water's stability field.) On the other hand, recall from Table 7.1 that S is found naturally in six different oxidation states. If S is only stable in two oxidation states, how can it be found in six? The answer lies in the difference between thermodynamics and kinetics. When S is in a reducing environment, it will tend to be in its thermodynamically stable form, namely, the −2 oxidation state. However, when this S is transported from its initial reducing environment to an oxidizing one, it suddenly finds itself in a thermodynamically unstable form. To reestablish thermodynamic equilibrium, the S will undergo a series of kinetic processes that sequentially transform it from its initial −2 oxidation state to an oxidation state of −1, and so on, until it finally reaches its thermodynamically stable state of +6. Similarly, when S is transported from an oxidizing environment to a reducing environment, kinetic processes will transform the S from the +6 oxidation state to its new stable state of −2. Because these kinetic processes do not occur instantaneously, S exists for finite time periods in its various intermediate oxidation states—hence the ubiquity of these oxidation states within the earth system.

7.3. THE KEY BIOGEOCHEMICAL REACTIONS OF THE S CYCLE

Although S is an important component of the biosphere, its most important biogeochemical reactions are those in which S remains in inorganic form and, by interchanging between its −2 and +6 oxidation states, acts as an electron donor and acceptor in a variety of key redox reactions. For example, in organic-rich, anoxic waters, *sulfate-reducing bacteria* can use sulfate ions (SO_4^{2-}) as an oxidizing agent to extract chemical energy from organic molecules via

$$SO_4^{2-} + 2\text{"}CH_2O\text{"} + 2H^+ \rightarrow (H_2S)_g + 2CO_2 + 2H_2O + \text{energy} \qquad (R7.1)$$

In shallow water systems, the gaseous H_2S produced from (R7.1) often escapes into the atmosphere and gives rise to the "rotten eggs" smell that is characteristic of salt marshes and tidal mudflats.

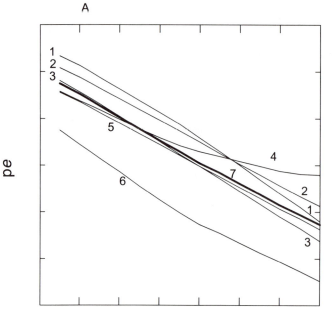

1. S(+4) - S(0)
2. S(+4) - S(-2)
3. S(+4) - S(0)
4. S(0) - S(-2)
5. S(-1) - S(-2)
6. S(+6) - S(+4)
7. S(+6) - S(-2)

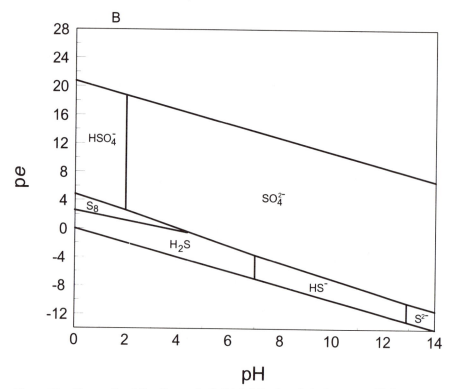

Figure 7.3. The pe–pH stability diagram for S. (**A**) Stability boundaries for seven oxidation-state couplets of S. (**B**) The pe–pH diagram for the stability region of water with all metastable boundaries eliminated. Note that only S compounds in the +6 p, 0 and −2 oxidation states appear as true stable states within the region where water is stable.

140

Before the advent of green-plant photosynthesis and the rise of atmospheric O_2, much of the H_2S produced via (R7.1) was probably converted back to SO_4^{2-} by photosynthetic *purple and green sulfur bacteria* via

$$(H_2S)_g + 2CO_2 + 2H_2O + h\nu \rightarrow SO_4^{2-} + 2\text{“}CH_2O\text{”} + 2H^+ \qquad \text{(R7.2)}$$

Under the more oxidizing conditions of the present-day atmosphere and hydrosphere, however, most of the H_2S is simply oxidized back to sulfate via a series of reactions whose net effect can be represented by the stoichiometric reaction (see Text Box 7.1)

$$(H_2S)_g + 2O_2 \rightarrow SO_4^{2-} + 2H^+ \qquad \text{(R7.3)}$$

It is interesting to note that while the sum of Reactions (R7.1) and (R7.2) constitute a closed cycle with no net production or loss of any compound, the same cannot be said of (R7.1) and (R7.3); that is,

$$SO_4^{2-} + 2\text{“}CH_2O\text{”} + 2H^{I} \rightarrow (H_2S)_g + 2CO_2 + 2H_2O + \text{energy} \qquad \text{(R7.1)}$$

$$(H_2S)_g + 2O_2 \rightarrow SO_4^{2-} + 2H^+ \qquad \text{(R7.3)}$$

Net: $\quad 2\text{“}CH_2O\text{”} + 2O_2 \rightarrow 2CO_2 + 2H_2O + \text{energy}$

TEXT BOX 7.1	ATMOSPHERIC CYCLE OF S DRIVEN BY OH RADICALS AND RELATED PHOTOCHEMICAL OXIDANTS

As discussed in Section 7.2, when reduced S compounds such as H_2S are emitted into the atmosphere they suddenly find themselves in a thermodynamically unstable state. Because of the relatively large abundance of molecular oxygen in the atmosphere, the +6 oxidation state of S is favored over its other more reduced states. As a result, any reduced S compounds in the atmosphere will tend to be oxidized. If this oxidation were to require direct reaction with molecular oxygen itself, the process of oxidation would be very slow indeed. Because of the strength of the O–O bond in molecular oxygen, it tends to react rather slowly with most compounds at ambient temperatures and pressures. Instead the atmosphere has an alternate pathway for oxidizing and ultimately cleansing itself of reduced S compounds (and, in fact, most other thermodynamically unstable species). This pathway involves the production of highly reactive photochemical oxidants such as hydroxyl radicals (OH) that react relatively rapidly with H_2S and other reduced compounds (see Figure 7.4), thereby causing their more rapid transformation to thermodynamically stable states. Because these thermodynamically favored states are more oxidized, the compounds thus formed tend to be more polar and therefore relatively soluble in water. For this reason, the products of atmospheric oxidation are easily removed from the atmosphere via rainout, washout, and other depositional processes. For example, in the case of H_2S, reaction with OH is believed to initiate a series of elementary reactions that lead to the eventual production of sulfur dioxide (SO_2). The SO_2, in turn, is oxidized to H_2SO_4, which forms sulfate-containing particles and is removed from the atmosphere, thereby returning the S to the natural water systems from whence it initially came.

At first we might be tempted to conclude from the above that Reactions (R7.1) and (R7.3) constitute a biogeochemical cycle that continually removes molecular oxygen (O_2) from the atmosphere. However, before coming to this conclusion we must ask ourselves where the organic molecules (i.e., "CH_2O") that appear on the left-hand side of the net reaction come from. In the present-day atmosphere, they of course come from green-plant photosynthesis

$$CO_2 + H_2O + h\nu \rightarrow \text{"}CH_2O\text{"} + O_2 \qquad (R1.1)$$

thus providing both the organic material and the O_2 which Reactions (R7.1) + (R7.3) require and, in the process, closing the cycle.

7.3.1. Formation and Weathering of Pyrite

The formation of hydrogen sulfide through anaerobic sulfate reduction and its subsequent oxidation forms a tightly closed cycle that, in large part, mimics the cycle of photosynthesis and respiration we have discussed earlier. Because of its similarity to the photosynthesis/respiration cycle and because, on a global scale, it processes only a fraction of the C actually produced by photosynthesis, the oxidation and reduction cycle of S represented by Reactions (R7.1) and (R7.3) is of limited interest from a global point of view. (On the other hand, it can be quite important in any number of anaerobic microenvironments.)

However, when anaerobic sulfate reduction occurs in ocean sediments containing hematite (Fe_2O_3), a cycle of major global importance can be initiated. Within these sediments, so-called *colorless bacteria* oxidize organic compounds by reducing the iron (Fe) in hematite as well as the S in the sulfate ions, and, in the process, produce an insoluble compound containing reduced S that can accumulate in the ocean sediment. The compound is pyrite (FeS_2), sometimes referred to as "fool's gold" because of its golden color (see Figure 7.5). The reactions that produce FeS_2 in the ocean sediment have the net stoichiometry of (R7.4):

$$8SO_4^{2-} + 2Fe_2O_3 + 15\text{"}CH_2O\text{"} + 16H^+$$
$$\rightarrow 4(FeS_2)_S + 15CO_2 + 23H_2O \qquad (R7.4)$$

The cycle initiated by (R7.4) is closed when the pyrite thus formed and accumulated in the ocean sediments is eventually brought to the earth's surface by tectonic processes and weathered. This weathering reaction is represented by (see Text Box 7.2)

$$4FeS_2 + 8H_2O + 15O_2 \rightarrow 2Fe_2O_3 + 8SO_4^{2-} + 16H^+ \qquad (R7.5)$$

The importance of this cycle arises from the fact that each time Reaction (R7.4) occurs, "CH_2O" is oxidized without the consumption of molecular oxygen (O_2). Because 1 mole of O_2 is produced in photosynthesis for each mole of "CH_2O" produced, it follows from (R7.5) that the burial of 1 mole of S as pyrite in an ocean sediment effectively represents a net source of 15/8 moles of O_2 to the atmosphere. Eventually, of course this O_2 is removed from the atmosphere when the pyrite is weathered via (R7.5). However, recall that the characteristic time for the uplift and weathering of ocean sediments is approximately 100 million years. Thus, any tem-

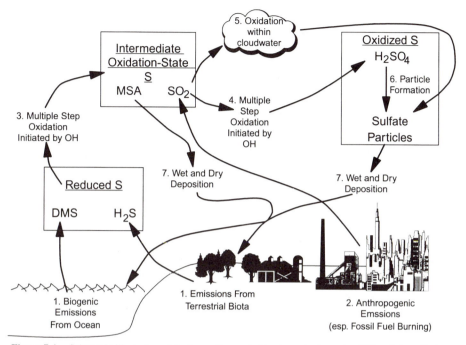

Figure 7.4. Schematic illustration of the key pathways in the atmospheric cycle of S involving: (1) the natural emissions of reduced S compounds such as H_2S from terrestrial biota and dimethylsulfide (CH_3SCH_3) from oceanic biota; (2) anthropogenic emissions of S compounds, principally SO_2; (3) the oxidation of reduced S compounds by OH and other photochemical oxidants leading to the production of intermediate oxidation state S compounds such as SO_2 and methanesulfonic acid (MSA); (4) the oxidation of these intermediate oxidation state compounds within the gas phase by OH-producing H_2SO_4 vapor; (5) the conversion of intermediate oxidation state compounds within liquid cloud droplets, which upon evaporation yield sulfate-containing particles; (6) the conversion of H_2SO_4 to sulfate-containing particles; and (7) the ultimate removal of S from the atmosphere by wet and dry deposition.

porary imbalance in the burial and weathering rates of pyrite has the potential to cause rather long-term perturbations to the O_2 content of the atmosphere. We will explore this possibility in greater detail in one of the numerical experiments presented later in this chapter.

7.3.2. Formation and Weathering of Gypsum

The other sulfur mineral of major global significance is gypsum ($CaSO_4 \cdot 2H_2O$), an evaporite deposit. Precipitation of gypsum, in its simplest form, might be expected to proceed via

$$Ca^{2+} + SO_4^{2-} + 2H_2O \rightarrow (CaSO_4 \cdot 2H_2O)_S \qquad (R7.6)$$

However, Reaction (R7.6) requires that the gypsum crystals nucleate homogeneously, a relatively slow process. It is far more common for the gypsum to nucleate heterogeneously by displacing carbonate ions (CO_3^{2-}) from calcite deposits; that is,

$$(CaCO_3)_S + H^+ + SO_4^{2-} + 2H_2O \rightarrow (CaSO_4 \cdot 2H_2O)_S + HCO_3^- \qquad (R7.7)$$

(a)

(b)

Figure 7.5. Panel A: Crystal of pyrite (FeS_2) from Logrono, Ambassaguas, Spain; approximately $2''$ in size. Panel B: Crystallized gold specimen from Breckenridge, Colorado, USA; approximately $^3/_4''$ in size. Pyrite is known as "fool's gold" because of its golden hue. Unlike gold, however, pyrite is a mineral found in sedimentary rock and typically has a layer-like structure. (Photos provided courtesy of the United States Geological Survey.)

TEXT BOX 7.2	WEATHERING OF PYRITE AND ACID MINE DRAINAGE A CAUSE OF LOCAL GROUNDWATER ACIDIFICATION

Pyrite is widely distributed in sediments and in the rocks that are formed from these sediments. As a result, mining operations, especially for coal and metal sulfide ores, often bring significant quantities of pyrite to the earth's surface. Because the pyrite is not of significant economic value, it was a common practice in years past to simply leave the pyrite at the mining site in so-called *mine tailings*. We know from our discussion in Section 7.2, however, that when pyrite is exposed to the atmosphere it will be weathered or oxidized via

$$4FeS_2 + 8H_2O + 15O_2 \rightarrow 2Fe_2O_3 + 8SO_4^{2-} + 16H^+ \qquad (R7.5)$$

resulting in the production of sulfuric acid. Eventually the sulfuric acid produced from this reaction will seep into the local groundwater, leading to a phenomenon known as *acid mine drainage*. Acid mine drainage has resulted in significant, and in some cases devastating, environmental damage from the ensuing acidification of the local groundwater system as well as the contamination from the trace metals liberated from the soils by this acidity. Careful isolation and storage of mine tailings or, even better, development of economic uses for the mine tailings represent two alternate strategies for dealing with this problem.

Interestingly, gypsum formation also has the potential to affect the oxygen content of the atmosphere. By removing carbonate from the ocean sediments and placing it back into the ocean reservoir where it can be used by photosynthesizing phytoplankton, Reaction (R7.7) can indirectly enhance production of atmospheric oxygen. As was the case for pyrite deposition, the cycle of gypsum deposition is closed when these sediments are brought to the earth's surface and weathered, via

$$(CaSO_4 \cdot 2H_2O)_S \rightarrow Ca^{2+} + SO_4^{2-} + 2\,H_2O \qquad (R7.8)$$

7.4. THE PREINDUSTRIAL GLOBAL S CYCLE

We begin our numerical analysis of the global biogeochemical cycle of S by constructing a steady-state representation of the cycle as it may have existed in preindustrial times—that is, in the absence of any anthropogenic perturbations. As illustrated in Figure 7.6, our model has five reservoirs; these include reservoirs for oxidized S sediments (i.e., gypsum), reduced S sediments (i.e., pyrite), S in the soils, S in the atmosphere, and S in the oceans. Following the same basic methodology outlined in the previous chapters, the contents of the various reservoirs and

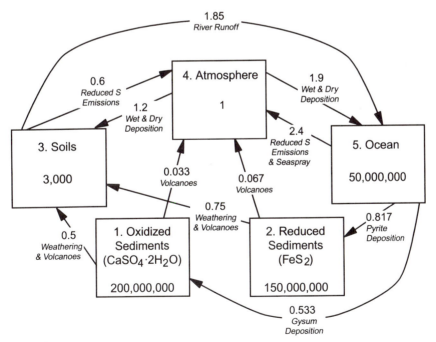

Figure 7.6. The five-reservoir model for the preindustrial, steady-state S cycle with separate reservoirs for oxidized and reduced S sediments. Reservoir amounts are given in units of Tmoles and fluxes in Tmoles year^{-1}. (Recall that 1 Tmole = 1×10^{12} moles.)

the fluxes between the reservoirs can be estimated. The values we will adopt here are indicated in Figure 7.6 and Table 7.2, respectively. Because we will want to eventually infer how perturbations to this cycle may affect the cycles of O and C, we find it convenient to use molar rather than mass units. Hence, reservoir contents are expressed in units of Tmoles of S and fluxes are expressed in Tmoles/ year.[1]

It is useful, at this point, to take note of the significant differences between the model we have adopted here for the S cycle and the models we had developed earlier for the P and C cycles. In those previous models, we had included reservoirs for the marine and terrestrial biospheres. Consideration of a marine biosphere had, in turn, necessitated a more detailed treatment of the ocean, with reservoirs for the surface ocean and the deep ocean. For the S cycle, however, no biospheric reservoirs are included. Moreover, unlike the P cycle but like the C cycle, our S cycle includes two distinct sedimentary reservoirs. These choices reflect our earlier discussions in this chapter. From a global perspective, the pivotal

TABLE 7.2
Preindustrial Fluxes of S for the Five-Reservoir Steady-State Model

Fluxes	Flux (Tmoles year^{-1})
From oxidized S sediments	
To soils ($F_{1\to3}$)	Weathering = 0.4
	Volcanoes = 0.1
	Total = 0.5
To atmosphere ($F_{1\to4}$)	Volcanoes = 0.033
From reduced S sediments	
To soils ($F_{2\to3}$)	Weathering = 0.6
	Volcanoes = 0.15
	Total = 0.75
To atmosphere ($F_{2\to4}$)	Volcanoes = 0.067
From soils	
To atmosphere ($F_{3\to4}$)	Reduced S emissions = 0.6
To ocean ($F_{3\to5}$)	River runoff = 1.85
From atmosphere	
To soils ($F_{4\to3}$)	Wet and dry deposition = 1.1
	Seasalt deposition = 0.1
	Total = 1.2
To ocean ($F_{4\to5}$)	Wet and dry deposition = 0.6
	Seasalt deposition = 1.3
	Total = 1.9
From ocean	
To atmosphere ($F_{5\to4}$)	Reduced S emissions = 1.0
	Seaspray = 1.4
	Total = 2.4
To oxidized sediments ($F_{5\to1}$)	Gypsum deposition = 0.533
To reduced sediments ($F_{5\to2}$)	Pyrite deposition = 0.817

[1] 1 Tmole = 1 teramole = 1×10^{12} moles.

aspects of the S cycle are the rates at which sedimentary material is formed and weathered, as well as the division of this sedimentary material between reduced and oxidized forms of S. We have designed our model to focus on these aspects of the cycle.

From the data in Figure 7.6 and Table 7.2, it is a straightforward task to construct the **K** matrix for the preindustrial S cycle. We obtain the following result:

$$\mathbf{K} = \begin{array}{ccccc} -2.67 \times 10^{-9} & 0.0 & 0.0 & 0.0 & 1.07 \times 10^{-8} \\ 0.0 & -5.45 \times 10^{-9} & 0.0 & 0.0 & 1.63 \times 10^{-8} \\ 2.50 \times 10^{-9} & 5.00 \times 10^{-9} & -8.17 \times 10^{-4} & 1.20 & 0.0 \\ 1.65 \times 10^{-10} & 4.47 \times 10^{-10} & 2.00 \times 10^{-4} & -3.10 & 4.80 \times 10^{-8} \\ 0.0 & 0.0 & 6.17 \times 10^{-4} & 1.90 & -7.50 \times 10^{-8} \end{array}$$

7.5. NUMERICAL EXPERIMENT #1—SIMULATION OF ENHANCED GYPSUM DEPOSITION DURING THE PERMIAN PERIOD

In the preindustrial, steady-state model for the S cycle developed in the previous section, the formation and weathering rates of the oxidized S sediments (i.e., gypsum) and the formation and weathering rates of the reduced S sediments (i.e., pyrite) are each, individually, in balance. As a result, the cycling of the S sediments in this model has no net effect on the O_2 content of the atmosphere. However, there is no a priori reason to assume that this has always been the case. In fact, examination of the geologic record using S isotopes as a diagnostic tool indicates that there were periods in the not-too-distant past when the rates of gypsum and pyrite formation were probably vastly different from today's rates (see Text Box 7.3). One such period is the Permian, when, over a 50-million-year time interval from about 275 to 225 million years ago, it appears that there was a great deal more deposition of gypsum than under the present geologic regime. (The Permian was also a period with a large amount of flooding of the oceans over the continents. The frequent transgression of oceanic waters over continental regions perhaps facilitated the formation of evaporite deposits, and, hence, the enhanced amounts of gypsum formation from this period.) In this section, we will use *BOXES* to explore the possible consequences of such a shift in the S sedimentation rates.

In the steady-state model, about 1.35 tmoles of S are laid down in ocean sediments each year, with about 1.5 moles of the S going to reduced or pyrite sediments for each mole of S that goes to oxidized or gypsum sediments (see Table 7.2). To create a cycle to mimic what may have occurred during the Permian Period, we assume no change in the total sedimentation rate of S (i.e., 1.35 tmoles/year), but rather a change in the ratio of reduced to oxidized sediments. Specifically, we will assume for our perturbed cycle that 1 mole of reduced S sediments is laid down for

TEXT BOX 7.3	ISOTOPIC COMPOSITION OF GYPSUM HOLDS KEY TO GEOLOGIC PAST

The most abundant isotope of S on the earth has an atomic weight of 32 (i.e., ^{32}S), and the second most abundant isotope is ^{34}S. On average, there are about 22.5 atoms of ^{32}S for each atom of ^{34}S. However, this ratio is not uniform; it is not uncommon to find variations in the ^{34}S-to-^{32}S ratio of as much as a few percent when analyzing various materials from the earth. Geologists have found that these small variations provide important clues to unraveling the earth's history. To understand how, we must first introduce some nomenclature. By convention, the relative amount or isotopic *fractionation* in a sample is expressed in terms of the number of parts per thousand (‰) the isotopic ratio deviates from a preestablished, standard ratio. For S fractionation, this is expressed as

$$\delta^{34}S \ (‰) = \{[(^{34}S/^{32}S)_{sample}]/[(^{34}S/^{32}S)_{standard}] - 1\}1000$$

where the standard ratio is taken to be that found in the Canyon Diablo Meteorite.

Now let us consider the S fractionation in the ocean and in gypsum and pyrite deposits. In the ocean today, the $\delta^{34}S$ is observed to be +20‰; in other words, oceanic S is .2% richer in ^{34}S or *heavier* than that of the Canyon Diablo Meteorite. When gypsum forms as an evaporite deposit in the ocean, it takes on essentially the same isotopic S fractionation as that of the ocean. However, the same is not true of pyrite. Because pyrite formation is mediated by microbes, and it is ever so slightly less costly, from an energy point of view, to reduce ^{32}S than ^{34}S, pyrite deposits at the bottom of the ocean tend to be *lighter* than the ocean; generally about 30‰ lighter. These facts suggest the following:

1. Analysis of the isotopic fractionation of gypsum deposits of known age can be used to infer the isotopic fractionation of the ocean over the geologic past.
2. Periods of preferential pyrite formation relative to today should have produced an isotopic fractionation in the ocean and, therefore, in the gypsum deposits formed at that time, which is relatively heavy (i.e., $\delta^{34}S > 20‰$).
3. Periods of preferential gypsum formation relative to today should have produced an isotopic fractionation in the ocean and, therefore, in the gypsum deposits formed at that time, which is relatively light (i.e., $\delta^{34}S < 20‰$).

With this information at hand, let's now look at the variations in $\delta^{34}S$ in gypsum deposits over the past several 100 million years illustrated in Figure 7.7. Interestingly, we see that the ratio has remained fairly constant at +20‰ over the past 100 million years or so, suggesting that the relative rates of gypsum and pyrite deposition must have been approximately constant over this time period. However, as we look further back into the past, the situation changes, with $\delta^{34}S$ fluctuating by ± 10‰ in a fairly irregular pattern over time scales of 50–100 million years. We can conclude that during these time periods the relative rates of gypsum and pyrite deposition must have been significantly different from those of today.

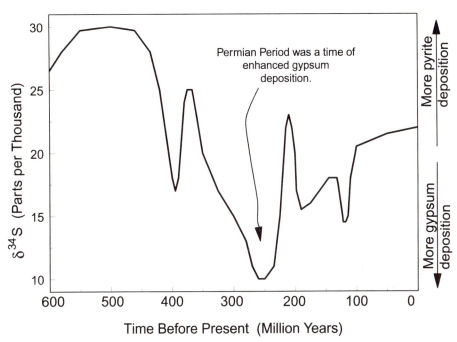

Figure 7.7.   Variations in sulfur isotope fractionation of gypsum deposits can be used to infer the relative rates of the formation of gypsum and pyrite sediments in the ocean. Periods with relatively high $\delta^{34}S$ were likely characterized by high rates of pyrite deposition, and periods with relatively low $\delta^{34}S$ (such as the Permian) were likely characterized by high rates of gypsum deposition. (After J. C. G. Walker, *Evolution of the Atmosphere*, Macmillan, New York, 1977.)

each 3 moles of oxidized S sediments. This perturbation gives rise to a slightly different **K** matrix:

$$\mathbf{K} = \begin{array}{ccccc} -2.67 \times 10^{-9} & 0.0 & 0.0 & 0.0 & \mathbf{2.02 \times 10^{-8}} \\ 0.0 & -5.45 \times 10^{-9} & 0.0 & 0.0 & \mathbf{6.75 \times 10^{-9}} \\ 2.50 \times 10^{-9} & 5.00 \times 10^{-9} & -8.17 \times 10^{-4} & 1.20 & 0.0 \\ 1.65 \times 10^{-10} & 4.47 \times 10^{-10} & 2.00 \times 10^{-4} & -3.10 & 4.80 \times 10^{-8} \\ 0.0 & 0.0 & 6.17 \times 10^{-4} & 1.90 & -7.50 \times 10^{-8} \end{array}$$

(where we have used a bold font to indicate the altered elements of the matrix).

Using this new matrix, the reservoir amounts in Figure 7.6 for our initial conditions, and running *BOXES* for a 50-million-year period (the approximate length of the Permian), we obtain the results illustrated in Figure 7.8 for the amounts of S in the reduced and oxidized sediment reservoirs. Because we made no changes to the elements of the **K** matrix that affect the weathering rates for the S sediments, we find, perhaps not very surprisingly, that changing the relative rates of reduced and oxidized S sedimentation produces a shift in the amounts of S in each of these reservoirs. For the values chosen in our specific experiment, we find that approximately 2×10^{19} moles of S shift from the reduced S reservoir to the oxidized S reservoir by the end of the 50-million-year period.

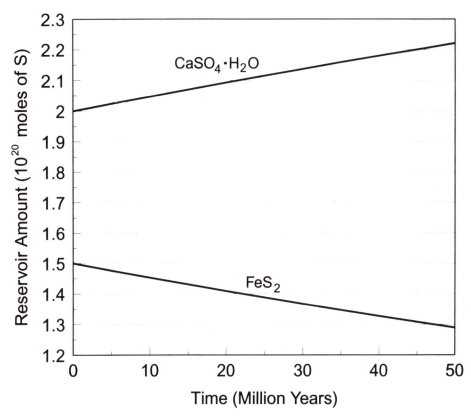

Figure 7.8. The time variation in the reservoirs of oxidized S sediments (CaSO$_4$·2H$_2$O) and reduced S sediments (FeS$_2$) when the rates of reduced S and oxidized S sedimentation are assumed to occur at a relative ratio of 1:3. Our calculations predict that about 2 × 10^{19} moles of S will be transferred from reduced sediments to the oxidized sediments after 50 million years as a result of the shift in the relative deposition rates. The interesting aspect of this calculation is to consider its implications for the abundance of atmospheric O$_2$.

The interesting aspect of this result is to consider what this change in the S content of the sediments implies for atmospheric O$_2$. Recall from Section 7.2 that for each mole of reduced S sediment oxidized, 15/8 moles of atmospheric O$_2$ are consumed. It follows, therefore, that the transfer of 2 × 10^{19} moles of S from reduced S sediments to oxidized S sediments over the 50-million-year period of our numerical experiment would also require the removal of roughly 4 × 10^{19} moles of O$_2$ from the atmosphere. The only problem with this scenario, and a rather significant one at that, is that there are only about 3.5 × 10^{19} moles of O$_2$ in the atmosphere. This implies that periods such as the Permian would have to have been essentially devoid of atmospheric O$_2$. But this was clearly not the case. The fossil record gives ample evidence, for example, for the existence of any number of respirating animals during the Permian.

One inference that has been made from this result is that there must be feedback mechanisms in the earth system that stabilize the O$_2$ content of the atmosphere and prevent it from being depleted by shifts in S sedimentation rates on geologic time scales. For example, Garrels, Lerman, and Mackenzie (*American Scientist*, **64,**

306–315, 1974), in their seminal paper entitled "Controls of Atmospheric O_2 and CO_2: Past, Present, and Future," have proposed a mechanism based on the fact that the deposition of gypsum via Reaction (R7.7) causes the dissolution of carbonate into the ocean. These authors argue that this extra carbonate is then available for uptake in photosynthesis, producing atmospheric O_2 in the process and thereby preventing its depletion. We will analyze the strength of this and other feedback mechanisms in Chapter 9, where we consider the coupling of the C, O, and S cycles in a more quantitative fashion.

7.6. NUMERICAL EXPERIMENT #2—EFFECT AND PERSISTENCE OF ANTHROPOGENIC PERTURBATIONS

Having considered the causes and consequences of perturbations to the S cycle that act on geologic time scales in Section 7.5, we now turn to a problem of more immediate relevance, namely, the extent to which the activities of our modern industrial society affect the S cycle and how rapidly these effects would dissipate if these perturbations were ceased. Anthropogenic activities currently perturb the S cycle in two principal ways. First, through mining and other land-use changes, they accelerate the rates at which both reduced and oxidized S sediments are weathered and transferred to the soils. Secondly, through the burning of fossil fuels, anthropogenic activities cause the direct transfer of reduced S from sediments to the atmosphere. Representative rates for each of these processes are listed Table 7.3.

As we did in the case of our numerical experiment on the C cycle, we adopt a hypothetical scenario that involves the imposition of anthropogenic perturbations to the steady-state cycle, the persistence of these perturbations for 130 years, and then the sudden halting of these perturbations. The treatment of the first 130 years of the scenario is carried out in Stage 1 of the *BOXES* simulation with the initial reservoir contents taken to be the values from the preindustrial, steady-state cycle and a **K** matrix that includes the anthropogenically driven fluxes listed in Table 7.3. This new, anthropogenically perturbed **K** matrix is given by

$$
\mathbf{K} =
\begin{array}{ccccc}
\mathbf{-4.67 \times 10^{-9}} & 0.0 & 0.0 & 0.0 & 1.07 \times 10^{-8} \\
0.0 & \mathbf{-2.28 \times 10^{-8}} & 0.0 & 0.0 & 1.63 \times 10^{-8} \\
\mathbf{4.50 \times 10^{-9}} & \mathbf{9.00 \times 10^{-9}} & -8.17 \times 10^{-4} & 1.20 & 0.0 \\
1.65 \times 10^{-10} & \mathbf{1.38 \times 10^{-8}} & 2.00 \times 10^{-4} & -3.10 & 4.80 \times 10^{-8} \\
0.0 & 0.0 & 6.17 \times 10^{-4} & 1.90 & -7.50 \times 10^{-8}
\end{array}
$$

where, as before, we have used a bold font to indicate the changed matrix elements.

TABLE 7.3
Sulfur Fluxes Caused by Anthropogenic Activities

Process	Flux (Tmoles year^{-1})
Transfer of oxidized S sediments to soils from mining (F_{13})	0.4
Transfer of reduced S sediments to soils from mining (F_{23})	0.6
Transfer of reduced S sediments to atmosphere from burning fossil-fuels (F_{24})	2.0

The treatment of the post-130-year period is carried out in Stage 2 of the *BOXES* simulation, using the reservoir contents obtained at time $t = 130$ years from Stage 1 as the initial reservoir contents and, because anthropogenic emissions have been assumed to cease in this stage, the preindustrial **K** matrix. By combining the first- and second-stage results, we obtain a record of the build-up and subsequent decay of the anthropogenically driven disturbance to the S cycle over this hypothetical 130-year period.

As illustrated in Figure 7.9, significant perturbations to the S contents of the atmosphere and the soils are predicted. While the atmospheric perturbation is predicted to be quite large (almost a factor of 2), the perturbation to the soils is more modest, reaching about 10% before anthropogenic emissions are hypothesized to stop at 130 years. As the reader might infer from inspection of the fluxes in Tables 7.2 and 7.3, the large increase in atmospheric S can be almost entirely attributed to the S emissions from the burning of fossil fuels, especially coal. Note also the very different response times for the atmosphere and soils, with the atmosphere only requiring about 1 year to adjust to the changing fluxes, while the soils' adjustment time is of the order of 1000 years. As the accompanying discussion describes (see Text Box 7.4), the rapid response of the atmosphere to S emissions presents policy makers with a perplexing conundrum.

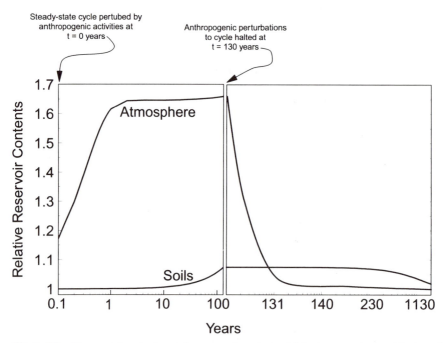

Figure 7.9. Time variations in the relative reservoir contents of S in the atmosphere and in the soils as a result of a hypothetical 130-year perturbation to the preindustrial, steady-state S cycle from mining and fossil fuel burning. The largest perturbation to the cycle is found to occur in the atmosphere. However, because of atmospheric S's short residence time, this perturbation will very quickly decay should anthropogenic emissions of S to the atmosphere be halted. The smaller perturbation to S concentrations in soils, although only of the order of 10%, is predicted to require about 1000 years to decay.

TEXT BOX 7.4	THE CONTROL OF S EMISSIONS FROM FOSSIL FUEL BURNING: A CONUNDRUM FOR POLICY MAKERS

Our simulations suggest that atmospheric S emissions from the burning of fossil fuels are having a major impact, causing as much as a factor of 1.5 to 2 increase in the element's average atmospheric concentration. The environmental effects of this increase are twofold and therein lies the conundrum for policy makers.

As illustrated in Figure 7.4, when S is emitted into the atmosphere, it is oxidized and transformed into sulfuric acid. This sulfuric acid, in turn, tends to nucleate and condense on aerosols, forming acidic sulfate particles. In large concentrations, these acidic particles pose a health threat to humans and, when removed from the atmosphere by wet and dry deposition, can lead to acid precipitation with its potential for harming sensitive ecosystems. For these reasons, policy makers would like to limit the emissions of S from fossil fuel burning.

But sulfate-containing particles also have an important climatic effect. Because of their optical properties, they tend to cool the climate by reflecting solar radiation back out to space before it can be absorbed by the atmosphere and the earth. In fact, some climate model simulations indicate that the buildup of sulfate-containing aerosols in the atmosphere since the industrial revolution has inadvertently, but nevertheless fortuitously, offset the warming effect of the concomitant buildup in the concentrations of greenhouse gases such as CO_2. Moreover, although the enhancements of CO_2 and other greenhouse gases will take decades to dissipate, the sulfate aerosol enhancements will dissipate almost immediately following a cessation of anthropogenic S emissions (compare Figures 6.10 and 7.8). Some scientists warn that if emission controls are implemented to reduce or eliminate S emissions from fossil fuel burning, the result will be a sudden climatic warming, as the full effect of the greenhouse gases in the absence of anthropogenic sulfate aerosols takes hold.

7.7. CONCLUSION

The rich diversity of environments in the earth system makes it possible for S to appear in oxidation states ranging from its most reduced state of -2 to its most oxidized state of $+6$. As a result, the biogeochemical cycle of S is centered about an array of redox reactions that cycle the element between these two extreme oxidation states. Because the amount of S that is cycled through these extreme oxidation states each year turns out to be relatively large, the biogeochemical cycle of S can have a major impact on the oxidative balance of the earth system itself, ultimately influencing the abundance of atmospheric oxygen and the cycling of organic C.

In addition to its important role in controlling the oxidative balance of the earth system, the global S cycle appears to be subject to significant perturbations. Perhaps most notable of these perturbations are those arising from anthropogenic activities. It seems fairly certain that S fluxes from the lithosphere to the soils and the atmosphere as a result of mining and fossil fuel burning have attained levels that challenge or even supersede natural transfer rates. As a result, S abundances in the both the atmosphere and soils have very likely been on the rise since the Industrial

Revolution. The potential consequences of these perturbations include degradation in local and regional air quality, acidification and toxification of local groundwater systems and ecosystems, and regional and global climate cooling. Determining the magnitude of these effects, as well as the response of the earth system to these effects, defines two areas of important study for the biogeochemist in the coming decades.

SUGGESTED READING

Charlson, R. J., J. E. Lovelock, M. O. Andreae, and S. G. Warren, Oceanic phytoplankton, atmospheric sulphur, cloud albedo, and climate, *Nature*, **326**, 655–661, 1987.

Charlson, R. J., J. Langner, and H. Rodhe, Sulphate aerosol and climate, *Nature*, **348**, 22, 1990.

Charlson, R. J., T. L. Anderson, and R. E. McDuff, The sulfur cycle, in *Global Biogeochemical Cycles*, edited by S. Butcher, R. J. Charlson, G. Orians, and G. V. Wolfe, Academic Press, New York, 1992.

Lawrence, M., An empirical analysis of the strength of the phytoplankton-dimethylsulfide-cloud-climate feedback cycle, *J. Geophys. Res.*, **98**, 20663–20673, 1993.

Taylor, K. E., and J. E. Penner, Response of the climate system to atmospheric aerosols and greenhouse gases, *Nature*, **369**, 734–737, 1994.

Wigley, T. M. L., Cloud reducing fossil-fuel emissions cause global warming?, *Nature*, **349**, 503–506, 1991.

PROBLEMS

1. One of the shortcomings in the five-box model we adopted for the S cycle in this chapter is the assumption of a uniform S abundance throughout the atmosphere. Because the atmospheric residence time for S is a few months and the time for a species to mix in the atmosphere is ~1 month (see Table 3.8 and/or Figure 3.15), it is likely that atmospheric S has significant gradients between regions where S is emitted and regions removed from these sources. Develop a six-box model for the S cycle that accounts for this variability by including separate reservoirs for atmospheric S over polluted continental regions and over clean, remote regions. Use *BOXES* to estimate the degree to which atmospheric S over each of these regions has increased as a result of anthropogenic activities.

2. In a paper published in *Nature* in 1987, Charlson, Lovelock, Andreae, and Warren proposed that the biogeochemical cycle of S acts as a natural climate stabilization mechanism within the earth system. In this mechanism, the production of dimethylsulfide (CH_3SCH_3) results in the emission of CH_3SCH_3 into the atmosphere and its oxidation to sulfate aerosol. Since sulfate aerosols tend to cool the earth's surface, these authors argued that surface temperatures would tend to be stabilized if the rate of CH_3SCH_3 production by ocean biota was enhanced by increasing temperatures and depressed by decreasing temperatures.

Using *BOXES* in a pseudo nonlinear mode (or some other suitable numerical scheme), estimate the strength of the relationships between atmospheric S and temperature, and also estimate temperature and CH_3SCH_3 production by ocean biota that would be required to stabilize atmospheric temperatures. Compare your results to those of Lawrence, M., *J. Geophys. Res.*, **98,** 20663–20673, 1993.

The Global
Nitrogen Cycle

"Nitrogen is 79 percent of the atmosphere, but it cannot be used directly by the large majority of living things. It must first be 'fixed'."

C. C. Delwiche, The Nitrogen Cycle, in *The Biosphere*,
W. H. Freeman, San Francisco, 1970.

8.1. INTRODUCTION

Nitrogen (N) lies between carbon and oxygen and above phosphorus in Group VA of the Periodic Table and has an atomic number of 7 and an atomic weight of 14.0067 (Figure 8.1). At an average terrestrial abundance of only 50 ppm by mass, N is the scarcest of the major nutrient elements. Nevertheless, the element is critical to all living organisms (Text Box 8.1) and, despite its low abundance, is the fourth most abundant element in organic matter, comprising roughly 0.3% of the mass of the biosphere. Of course N is found in even greater abundance in another of the earth's spheres, the atmosphere, where nitrogen is almost 80% of the mass.

Ironically, although most green plants exist within a veritable sea of atmospheric nitrogen, they must continuously struggle to get the nitrogen they need to carry out their metabolic functions, and the productivity of many ecosystems is ultimately limited by the ability of photosynthesizers within these ecosystems to assimilate N. The reason for this paradox arises from the fact that atmospheric nitrogen is in a highly stable chemical form (i.e., N_2) that cannot be used by most organisms. Before the N in the atmosphere can be assimilated by green plants, the triple bond that holds N_2 together must first be cleaved forming *fixed nitrogen*—that is, N not bonded to another N atom. This process requires energy, and the biosphere has devised a variety of pathways for producing fixed nitrogen and preventing its conversion back to the molecular form. To understand how this is accomplished, we must first turn to a discussion of N's redox chemistry.

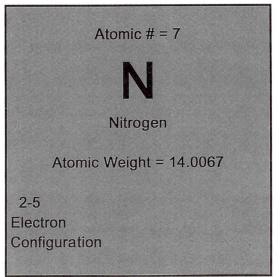

8.2. N REDOX CHEMISTRY

Like phosphorus, nitrogen has five valence electrons and thus can have oxidation states ranging from $+5$ to -3. However, unlike P, which is only found in one oxidation state, N occurs in nature in six of its eight possible oxidation states and in all three phases. A listing of the N compounds commonly found in each of the earth's spheres is presented in Table 8.1. Two notable features of this list are the following: (i) Virtually all N found in organic matter is associated with an amino group, and, as a result, organic N is nearly always in the -3 oxidation state; and (ii) there are no N-containing minerals of any significance in the lithosphere, and, hence, N in the solid earth occurs as partially decayed organic matter in soils and ocean sediments, dissolved nitrates, nitrites, and ammonia in soil pore water, or as ammonium ions trapped within silicate minerals.

Figure 8.2 summarizes the thermodynamic relationships between N's oxidation states. On the basis of these data and those listed in the appendix, a series of linear equations describing the stability boundaries between N's various oxidation states can be derived (Figure 8.3A). By eliminating the metastable boundaries and superimposing the stability boundaries for water, we obtain our final pe–pH stability diagram for N illustrated in Figure 8.3B. The diagram indicates the existence of three stable oxidation states for N: -3, 0, and $+5$. Although N_2 dominates over most of the stability region of water, NO_3^- is the most stable state in the region close to the H_2O/O_2 stability line where most of the environments encountered in the earth system tend to cluster (see Figure 2.7). This facet of N redox chemistry makes the N cycle similar to the cycle of C. Like organic C, N found in organic matter (i.e., in the -3 oxidation state) is usually thermodynamically unstable, and thus the production, assimilation, and maintenance of nitrogen in organic matter represents an energy drain for the biosphere.

The basic structural unit from which most proteins are formed is the amino acid, and most amino acids contain N in the form of an amino group (NH_2). Amino acids also contain a carboxyl group (COOH), a hydrogen atom, and an additional side chain or organic radical group (R). In un-ionized form, these amino acids are typically represented by

Actually, an amino acid always has one or more formally charged atoms, depending on its structure and pH. When R contains no additional amino or carboxyl groups, an amino acid is a diprotic acid whose form varies with pH (see Chapter 2). For example,

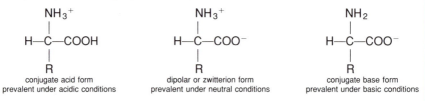

In nature, there are 20 different amino acids, each with a distinct side chain. For example, the simplest of the amino acids is *glycine*, which has a hydrogen atom for a side chain. Next in complexity is alanine, with R = CH_3. Two amino acids (cysteine and methionine) contain S in their side chains.

When amino acids combine, they form a chemical bond between the N atom of the amino group of one acid and the C atom of the carboxyl group of the other:

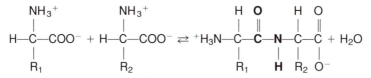

The bond that forms between the C and N atom is referred to as a *peptide bond* (indicated by bold type).

A polypeptide chain is simply a series of these peptide bonds strung together in a linear chainlike structure:

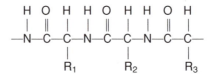

In general, polypeptide chains can be quite complex involving literally hundreds of peptide bonds and various combinations of side chains. Proteins, the stuff of all living matter, consist of one or more of these polypeptide chains. (Multiple-chained proteins are often held together by disulfide bonds between the chains.)

TABLE 8.1

Naturally Occurring N Compounds and Their Oxidation States

| Oxidation state | Atmosphere | | Hydrosphere | Lithosphere | | Biosphere |
	Gas	Particulate		Soils	Rocks	
-3	NH_3 $RNH_2{}^a$	NH_4HSO_4, $(NH_4)_2SO_4$, NH_4NO_3, NH_4Cl	NH_3, $NH_4{}^+$, NH_2CONH_2, amino acids, humic material	$NH_4{}^+$, amino acids, humic material	Humic material, $NH_4{}^+$ in silicates	Proteins DNA RNA
0	N_2	—	N_2	—	—	—
$+1$	N_2O	—	N_2O	—	—	—
$+2$	NO	—	—	—	—	—
$+3$	HNO_2	—	HNO_2, $NO_2{}^-$	$NO_2{}^-$	—	—
$+4$	NO_2	—	—	—	—	—
$+5$	HNO_3, N_2O_5	HNO_3, NH_4NO_3, $NaNO_3$, $Ca(NO_3)_2$	HNO_3, $NO_3{}^-$	$NO_3{}^-$	—	—

^{a}"R" is used to denote an organic radical (e.g., CH_3).

On the other hand, the fact that the most extreme N oxidation states (-3 and $+5$) are stable in the earth's more oxidizing and reducing environments makes the N cycle similar to the S cycle. Organisms able to exist within these extreme environments can extract energy from N-containing compounds transported into these environments by catalyzing their reduction or oxidation. These processes and their relevance to the global N cycle are examined in the next section.

$$NO_3^- \xleftarrow{(13.03)} (NO_2)_g \xleftarrow{(15.61)} NO_2^- \xleftarrow{(19.77)} (NO)_g \xleftarrow{(27.11)} \tfrac{1}{2}(N_2O)_g$$
$$[+5] \qquad\qquad [+4] \qquad\qquad [+3] \qquad\qquad [+2] \qquad\qquad [+1]$$

$$\tfrac{1}{2}(N_2O)_g \xleftarrow{(29.63)} \tfrac{1}{2}(N_2)_g \xleftarrow{(13.92)} NH_4^+$$
$$[+1] \qquad\qquad [0] \qquad\qquad [-3]$$

Figure 8.2. The oxidation states of N. The number in square brackets beneath each N-bearing species indicates the oxidation state of N within the compound, and the number in parentheses between each pair of compounds indicates the log of the equilibrium constant for the redox half-reaction between the two compounds.

8.3. THE KEY BIOGEOCHEMICAL REACTIONS OF THE N CYCLE

In the previous section we found that the N has three stable oxidation states within the stability region of water. It turns out that the biogeochemical cycling of the element is dominated by processes that shuttle N between these three oxidation states. These processes are schematically illustrated in Figure 8.4 and are briefly described below.

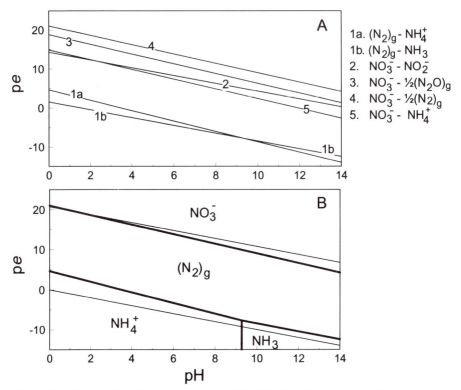

Figure 8.3. The pe–pH stability diagram for N. **(A)** Stability boundaries for six oxidation-state couplets of N. **(B)** The pe–pH diagram with all metastable boundaries eliminated (heavy solid lines) and the stability boundaries for H_2O superimposed (thin solid lines). Note that only N compounds in the +5, 0, and −3 oxidation states appear as true stable states within the region where water is stable. (*Note:* All stability boundaries were determined assuming 1 atm partial pressures for N-containing gases and 1 M concentrations for N-containing solutes.)

8.3.1. Nitrogen Fixation—Biotic and Abiotic

As noted earlier, most autotrophs are not able to directly assimilate atmospheric nitrogen. Because of the strength of the triple bond that holds its two N atoms together (225 Kcal/mole), N_2 is a relatively inert gas.[1] Fortunately, however, the earth system has developed a number of pathways for breaking the N–N bond in atmospheric nitrogen and producing fixed nitrogen that is then easily incorporated into organic matter. These pathways include *biotic nitrogen fixation*, where atmospheric nitrogen is converted to fixed nitrogen (usually ammonia) by living organisms, and *abiotic nitrogen fixation*, where the fixation occurs without biological intervention.

Biotic nitrogen fixation is carried out by autotrophs or by organisms in *symbiosis* with autotrophs, and hence it results in the production of organic N. Stoichio-

---

[1]In fact, when Antoine Lavoisier isolated the N_2 from air in the eighteenth century, he was so impressed with its inert properties that he called it *azote* from the Greek word meaning "without life" (M. E. Weeks and H. M. Leicester, *Discovery of the Elements, Journal of Chemical Education*, Easton, PA, 896 pages, 1968).

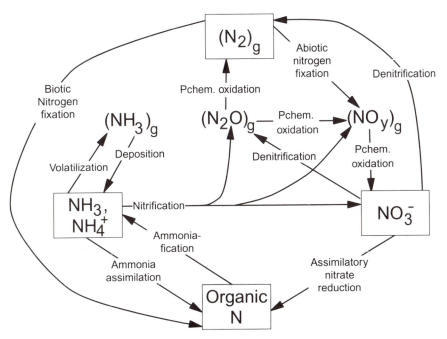

Figure 8.4. The key biogeochemical pathways of the nitrogen cycle. (*Note:* "Pchem oxidation" signifies photochemical oxidation via atmospheric chemical reactions.)

metrically, we can represent this process as a sequence of two steps. The first involves the production of ammonia from N_2 and water, that is,

$$N_2 + 5H_2O \rightarrow 2NH_4^+ + 2OH^- + 1\tfrac{1}{2}O_2 \qquad (R8.1)$$

and the second involves the assimilation of the ammonia into organic matter, as described in the next section.

The ability to fix nitrogen biologically is limited to a number of highly specialized organisms that contain the enzyme *nitrogenase*. This enzyme acts as a catalyst, facilitating the energy-intensive first step in nitrogen fixation, the cleavage of the N–N bond. Nitrogen-fixing organisms include symbiotic bacteria (e.g., *Rhizobium* and *Frankia*) and asymbiotic or free-living bacteria in soils and blue-green algae (or cyanobacteria) in surface waters.

In nature, abiotic nitrogen fixation is most often accomplished by a rapid injection of energy (often as heat or radiation) into air. The high temperatures that result from the energy injection cause the otherwise stable N_2 molecule to dissociate, and a small fraction of the N atoms thus produced react with atmospheric oxygen to form gaseous nitric oxide (NO). Thus, in contrast to biotic fixation which produces organic N, abiotic fixation in nature produces atmospheric nitrogen oxide gas and can be represented by the following overall stoichiometry:

$$(N_2)_g + (O_2)_g \rightarrow 2(NO)_g \qquad (R8.2)$$

What happens to the NO from abiotic fixation once it is released into the atmosphere? As it turns out, a lot, but that comes later, so keep reading.

Abiotic nitrogen fixation is accomplished by a host of energy-intensive, natural phenomena including lightning and forest fires. However, humankind is also a major and growing contributor to the earth's reservoir of fixed nitrogen, most notably through the production of nitrogen fertilizers and the burning of fossil fuels. Later in this chapter we will use *BOXES* to determine if this anthropogenic source of fixed nitrogen is perturbing the global biogeochemical cycle of N.

8.3.2. Ammonia Assimilation or Photosynthesis

Fixed nitrogen is most easily incorporated into biomass via *ammonia assimilation* in which dissolved ammonia is taken up by autotrophs and converted to organic nitrogen. Because this process is one portion of the overall process by which green plants build complex biomolecules, we can represent the stoichiometry of ammonia assimilation using the reactions derived in Chapter 3 for photosynthesis. Thus, for the ocean the stoichiometry of ammonia assimilation is given by

$$106CO_2 + 64H_2O + 16NH_3 + H_3PO_4 + h\nu$$
$$\rightarrow C_{106}H_{179}O_{68}N_{16}P + 106O_2 \qquad (R3.3)$$

and for the continents it is given by

$$830CO_2 + 600H_2O + 9NH_3 + H_3PO_4 + h\nu$$
$$\rightarrow C_{830}H_{1230}O_{604}N_9P + 830O_2 \qquad (R3.4)$$

8.3.3. Assimilatory Nitrate Reduction

Although not as energetically efficient as ammonia assimilation, the nitrate ion can also be used directly with most autotrophs to build complex biomolecules. In doing this, the organisms must first remove the nitrate from the environment and reduce it to ammonia; that is,

$$NO_3^- + H_2O + 2H^+ \rightarrow NH_4^+ + 2(O_2)_g \qquad (R8.3)$$

Once this is accomplished, the ammonium ion can then be used to produce proteins in the same manner as in ammonium assimilation via Reactions (R3.3) and (R3.4). Because the process involves nitrate reduction as well as assimilation, it is referred to as *assimilatory nitrate reduction*.

8.3.4. Ammonification or Mineralization

Just as the assembly of organic matter requires ammonia (either directly through ammonia assimilation or indirectly through assimilatory nitrate reduction), the breakdown of organic matter during respiration and decay leads to the release of ammonia. For this reason, respiration and decay is often referred to as *ammonification* when dealing with the N cycle. The overall stoichiometry of ammonification is of course represented by the reverse of Reaction (R3.3) in the ocean and Reaction (R3.4) on land.

8.3.5. Nitrification

The release of ammonia into oxidizing environments presents an opportunity for energy-hungry organisms, because ammonia is thermodynamically unstable in these environments. So-called *nitrifying bacteria* take advantage of this opportunity by catalyzing the oxidation of the reduced nitrogen and extracting the energy released from the oxidation to carry out their metabolic processes. Nitrifying bacteria fall into two categories: (i) *Nitrosomonas* bacteria, which catalyze the conversion of ammonium to nitrite via Reaction (R8.4),

$$NH_4^+ + \tfrac{3}{2}(O_2)_g \rightarrow NO_2^- + 2H^+ + H_2O \qquad \text{(R8.4)}$$

and (ii) *nitrobacteria*, which oxidize nitrite to nitrate via Reaction (R8.5),

$$NO_2^- + \tfrac{1}{2}(O_2)_g \rightarrow NO_3^- \qquad \text{(R8.5)}$$

Thus, when carried to completion, *nitrification* converts N from the -3 oxidation state to the $+5$ oxidation state. However, the process of nitrification is not 100% efficient, and generally a small fraction of the N being nitrified is converted to less-oxidized, gaseous species such as NO, NO_2, and N_2O and lost to the atmosphere. The production of these species most likely occurs as a result of side reactions involving nitrite ions produced by Reaction (R8.4) which probably have overall stoichiometries of

$$2H^+ + 2NO_2^- \rightarrow (NO)_g + (NO_2)_g + H_2O \qquad \text{(R8.6)}$$

and

$$2H^+ + 2NO_2^- \rightarrow (N_2O)_g + H_2O + (O_2)_g \qquad \text{(R8.7)}$$

8.3.6. Ammonia Volatilization

Ammonia is a species with a relatively high volatility and, when present in dissolved form as an ammonium ion (NH_4^+), can be converted to gaseous form, especially under alkaline or basic conditions, via

$$NH_4^+ + OH^- \rightarrow (NH_3)_g + H_2O \qquad \text{(R8.8)}$$

When Reaction (R8.8) occurs in a natural environment such as the pore water of soils (or the ocean), the ammonia gas thus produced can diffuse into the atmosphere. This reaction thus provides a pathway for transferring fixed nitrogen in the -3 oxidation state from the lithosphere (or ocean) to the atmosphere and is commonly referred to as *ammonia volatilization*.

8.3.7. Atmospheric Chemistry

As a result of abiotic nitrogen fixation, nitrification, and, as we shall see in the next section, denitrification, the gases NO, NO_2, and N_2O are generated and released into the atmosphere. As a result of ammonia volatilization, the gas NH_3 is released into the atmosphere. What happens to these gases once they are in the atmosphere and what, if any, effects do they have on the environment? These are the questions that are typically addressed in the discipline of atmospheric chemistry. In the sections below (and in Table 8.2) we summarize the answers that the field of atmospheric chemistry has been able to provide.

8.3.7.1. Atmospheric N_2O

N_2O, also known as *laughing gas* because of its anesthetizing properties in humans, is a linear molecule with an oxygen atom at one end, a nitrogen atom on the other, and a N atom in the middle ($N = N = O$). In large part because of the stability of the chemical bond that connects its two N atoms, N_2O is the least reactive of the nitrogen oxide gases. It is essentially inert in the troposphere, and its removal from the atmosphere requires that it first be transported into the stratosphere, where *photochemical reactions* triggered by ultraviolet radiation from the sun consume it. Because the rate of transport from the troposphere to the stratosphere is slow, N_2O has a relatively long atmospheric residence time of about 150 years.

The vast majority of the N in the atmosphere as N_2O is converted to molecular nitrogen, either through direct *photolysis* by ultraviolet photons via[2]

$$N_2O + h\nu \text{ (ultraviolet)} \rightarrow N_2 + O \qquad \text{(R8.9)}$$

or through reaction with electronically excited O atoms (O*) via

$$N_2O + O^* \rightarrow N_2 + O_2 \qquad \text{(R8.10a)}$$

The O* atom in Reaction (R8.10a) is, in turn, produced from photochemical reactions, primarily the photolysis of molecular oxygen and ozone by ultraviolet photons. However, N_2 is not produced every time N_2O reacts with O*; about half the time the reaction takes an alternate branch and yields NO instead of N_2 via

$$N_2O + O^* \rightarrow 2NO \qquad \text{(R8.10b)}$$

It is estimated that Reaction (R8.10b) is responsible for converting about 5% of the N_2O in the atmospheric into stratospheric NO, thus giving rise to one of N_2O's most important environmental roles: namely, exerting a controlling influence over the stratospheric ozone layer by providing the major source for stratospheric NO. (As indicated in Table 8.2, N_2O is also a greenhouse gas and thus its presence in the atmosphere tends to warm the climate.)

The processes that control the atmospheric abundance of N_2O have come under increased scrutiny by atmospheric chemists over the past decade or so for two reasons: (i) the recognition of N_2O's important role as a greenhouse gas and a source of stratospheric NO and (ii) the collection of atmospheric and ice core data documenting a slow and steady increase in atmospheric N_2O concentrations since the Industrial Revolution. These data indicate that N_2O concentrations have increased from about 275 ppbv before the Industrial Revolution to its present-day level of about 310 ppbv, an approximate 13% increase. The data also indicate a continuing increase in atmospheric N_2O today, at a rate of about 0.3% per year. The fact that N_2O concentrations have increased since the Industrial Revolution suggests that hu-

---

[2]Because we are dealing with atmospheric reactions, the species appearing in this and other reactions in this section are, unless otherwise noted, gases. For simplicity, we will not use the subscript "g" to indicate this fact.

TABLE 8.2
The Environmental Effects of N-Containing Trace Gases

N_2O	NO_y	NH_3
1. Global Warming N_2O is a greenhouse gas. The approximate 13% increase in N_2O concentrations since the Industrial Revolution may be contributing to global warming.	**1. Atmospheric Oxidation** NO_x helps determine the tropospheric concentration of OH radicals which, in turn, control the rate at which a variety of pollutants are oxidized and removed from the atmosphere. Increasing emissions of NO_x may be altering the oxidizing capacity of the atmosphere.	**1. New Particle Formation** NH_3, in conjunction with water and sulfuric acid vapor, acts to form new particles in the atmosphere via a process known as *termolecular nucleation*. These new particles may act to cool the climate thereby offsetting in part climate warming from greenhouse gases (see relevant discussion in Chapter 7).
2. Stratospheric Ozone The oxidation of N_2O in the stratosphere produces NO_x. NO_x in the stratosphere can either contribute to or prevent the catalytic removal of stratospheric ozone depending upon the stratospheric abundance of reactive chlorine compounds. Increasing N_2O concentrations may therefore be contributing to changes in the abundance of stratospheric ozone.	**2. Photochemical Smog** NO_x, in the presence of volatile organic compounds (i.e., hydrocarbons) and sunlight, catalyze, the generation of low-altitude ozone and photochemical smog. High NO_x concentrations in and around urban areas are partly responsible for deteriorating air quality on urban and regional scales.	**2. Acid Neutralization** NH_3 is one of only a handful of basic atmospheric species, and therefore it plays a critical role in neutralizing the acids in aerosols and rainwater.
	3. Acid Rain The oxidation of NO_x produces nitric acid vapor (HNO_3). The removal of this species by wet and dry deposition affects the acidity of precipitation and may be contributing to the phenomenon known as "acid rain."	**3. Source of NO_x** The oxidation of atmospheric NH_3 may represent a source (or a sink) of atmospheric NO_x.

Suggested reading for more information on the chemistry and effects of N-containing trace gases:
Chameides, W. L., and D. D. Davis, Chemistry in the troposphere, *Chemical and Engineering News*, **60**, 38–52, 1982.
Seinfeld, J., et al., *Rethinking the Ozone Problem in Urban and Regional Air Pollution*, National Academy Press, Washington, D.C., 500 pages, 1991.
Warneck, P., *Chemistry of the Natural Atmosphere*, International Geophysical Series, Vol. 41, Academic Press, San Diego, 1988.
Wayne, R. P., *Chemistry of Atmospheres*, Oxford University Press, Oxford, 447 pages, 1993.

mankind's growing population and concomitant intensification in agricultural and industrial activities have caused, or at least contributed to, this increase. Can we take this supposition one step further by developing a model that links humankind's perturbations to the global N cycle to the observed increase in N_2O? We will use *BOXES* to investigate this question later in this chapter.

8.3.7.2. Atmospheric NO_y

The term NO_y is used by atmospheric chemists to denote the family of nitrogen oxide gases in which N is not bonded to another N atom. (N_2O is a nitrogen oxide gas, but is not considered part of the NO_y family because the molecule contains an N–N bond.) NO_y compounds are, in turn, often divided into two subfamilies: (1) NO_x, which encompasses the sum of NO and NO_2, the two nitrogen oxides emitted directly into the atmosphere; and (2) NO_z, which encompasses all the NO_y compounds that are produced from the photochemical oxidation of NO_x (e.g., NO_3, N_2O_5, HNO_3, and organic nitrates). Thus,

$$NO_y = NO_x + NO_z$$

$$= (NO + NO_2) + (NO_3 + N_2O_5 + HONO + HNO_3 + RNO_3 + \cdots) \qquad (8.1)$$

where R represents an organic radical or molecular fragment.

A schematic diagram of the NO_y compounds and the photochemical reactions that couple them together is presented in Figure 8.5. In terms of its general overall

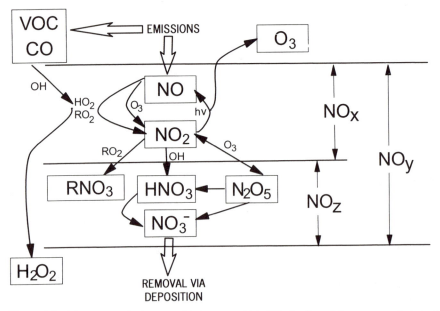

Figure 8.5. The compounds that comprise the atmospheric NO_y family and the processes that couple them together. These processes include the following: the emissions of NO_x ($=NO + NO_2$); the photochemical reactions that cycle NO and NO_2 and convert NO_x into NO_z compounds; and the removal of NO_y from the atmosphere by dry and wet deposition of HNO_3 and particulate NO_3^-. Note that the photochemistry of NO_y involves the oxidation of volatile organic compounds (VOC) and carbon monoxide (CO) and the generation of ozone (O_3) and hydrogen peroxide (H_2O_2) in the presence of OH, HO_2, and RO_2 radicals.

features, the photochemistry of NO_y is similar to that of the sulfur oxides, discussed in Chapter 7 and schematically illustrated in Figure 7.5. Like that of SO_2, NO_x oxidation is initiated by reaction with OH (in this case, the reaction of OH with NO_2) and results in the production of an acid (in this case, HNO_3 vapor). Moreover, like the sulfur oxides, the removal of NO_y from the atmosphere occurs primarily through the dry and wet deposition of its most oxidized state (in this case, nitrate).

On the other hand, there are some important dissimilarities between the atmospheric chemistry of the sulfur and nitrogen oxides. For example, sulfur oxide chemistry is essentially a linear process with chemical reactions always moving the S to higher and higher oxidation states. Although the overall tendency of NO_y chemistry is also toward higher oxidation states, the chemistry has embedded within it cyclic reaction sequences that convert N back and forth between more oxidized and less oxidized species. Most notable of these cyclic reaction sequences are those that couple NO and NO_2, because they give rise to photochemical mechanisms that catalytically produce and destroy ozone in the troposphere and stratosphere and make NO_y central to the chemical state of the atmosphere (see Text Box 8.2).

Because atmospheric NO_y compounds are reactive, they have relatively short atmospheric residence times and are generally removed from the atmosphere within a few days to a week or two after being emitted as NO_x. This short residence time prevents these compounds from being well-mixed in the atmosphere (recall from Chapter 3 that atmospheric mixing times are of the order of 1–2 months), and, as a result, the concentration of atmospheric NO_y is quite variable. NO_y concentrations typically range from less than 0.1 ppbv in pristine environments of the lower troposphere (which are far removed from significant NO_x sources) to levels approaching 1 ppmv in the polluted urban cores of the world with their large sources of NO_x from fossil fuel burning. The fact that the highest concentrations of NO_y are usually found in regions with large anthropogenic NO_x sources suggests that, as in the case of N_2O, humankind is altering the abundance of atmospheric NO_y. We will address this possibility using *BOXES* in Section 8.5.

8.3.7.3. Atmospheric NH_3

In contrast to the other N-containing gases emitted into the atmosphere, ammonia has a relatively high solubility in water. As a result, most of the ammonia in the atmosphere is removed by wet and dry deposition of ammonium ions, thus closing the atmospheric cycle initiated by ammonia volatilization. Although this cycle does not involve a change in the oxidation state of the N, it nevertheless represents a significant aspect of the N cycle. Because the location of the volatilization and deposition are rarely identical, the two processes act to redistribute fixed nitrogen within the soil and oceanic reservoirs. (It is perhaps not surprising therefore that farmers, who must expend significant resources applying N-fertilizers to their soils, attempt to limit the loss of ammonia from their soils via volatilization by manipulating the manner in which fertilizer is applied, the timing of irrigation, and so forth.)

Although most of the ammonia in the atmosphere is removed by wet and dry deposition, a small fraction (approximately 5–10%) is oxidized by OH via Reaction (R8.16):

$$NH_3 + OH \rightarrow NH_2 + H_2O \quad \text{(R8.16)}$$

THE PHOTOCHEMICAL CYCLING OF NO_x PROVIDES CRITICAL CATALYTIC PRODUCTION AND DESTRUCTION MECHANISMS FOR OZONE

The two species that comprise NO_x are coupled together by a number of reaction sequences that rapidly cycle N back and forth between NO and NO_2. Most often this cycling is accomplished by the sequence of Reactions (R.8.11), (R8.12), and (R8.13):

$$NO + O_3 \rightarrow NO_2 + O_2 \qquad (R8.11)$$

$$NO_2 + h\nu \rightarrow NO + O \qquad (R8.12)$$

$$O + O_2 + M \rightarrow O_3 + M \qquad (where\ M = N_2\ or\ O_2) \qquad (R8.13)$$

This reaction sequence is a *null* cycle—that is, a cycle where there is no net production or destruction of any species (i.e., every species that appears on the left-hand side of one reaction appears on the right-hand side of another). The reaction sequence is nevertheless quite important. These reactions occur in rapid succession and cycle the NO_x compounds on relatively short time scales (i.e., minutes). Because of the short cycling time, the sequence tends to establish a so-called *photostationary state*, in which the rates of each reaction are equal and the relative concentrations of NO and NO_2 are fixed for a given ozone concentration.

In the troposphere and lower stratosphere, an alternate reaction sequence occurs occasionally when NO reacts with hydroperoxyl radicals (HO_2) instead of ozone via Reaction (R8.14), giving rise to

$$NO + HO_2 \rightarrow NO_2 + OH \qquad (R8.14)$$

$$NO_2 + h\nu \rightarrow NO + O \qquad (R8.12)$$

$$O + O_2 + M \rightarrow O_3 + M \qquad (R8.13)$$

$$\overline{Net:\ \ HO_2 + O_2 \rightarrow OH + O_3}$$

In contrast to the earlier sequence, there is a net chemical change in this sequence: HO_2 is converted to OH, and O_3 is produced. Because NO_x is neither produced or destroyed, we say that the sequence of Reactions (R8.14), (R8.15), and (R8.13) is a *catalytic cycle*. This and similar reaction sequences involving organic peroxy radicals are now believed to provide an important pathway for generating ozone and OH in both pristine and polluted regions of the troposphere.

In the stratosphere, where atomic oxygen concentrations are sufficiently large, an alternate reaction sequence occurs which provides an important *catalytic destruction cycle* for ozone:

$$NO + O_3 \rightarrow NO_2 + O_2 \qquad (R8.11)$$

$$O_3 + h\nu \rightarrow O_2 + O \qquad (R8.15)$$

$$NO_2 + O \rightarrow NO + O_2 \qquad (R8.12)$$

$$\overline{Net:\ \ 2O_3 + h\nu \rightarrow 3O_2}$$

It is largely through these catalytic cycles that the nitrogen oxides, with their minute atmospheric abundance, end up playing such a critical role in the chemistry of the atmosphere and, by extension, the quality of our environment.

resulting in the production of NH_2, a free radical with the N atom in a -2 oxidation state.

There are a number of possible chemical pathways that may be taken by the NH_2 radicals thus produced. These include: (i) a reaction with NO_2 that results in the production of N_2O, (ii) a reaction with NO that results in the production of N_2, and (iii) a reaction with O_2 that results in the production of NO. However, because of uncertainties in the relevant chemical kinetics, it is difficult to reliably estimate the relative importance of these pathways, and, as a result, there is a good deal of uncertainty in the fate of the N that undergoes Reaction (8.16).

8.3.8. Denitrification

The final step in the biogeochemical cycling of N is *denitrification*, the process by which fixed nitrogen is converted back to its molecular form and returned to the atmosphere, thereby closing the cycle that began with nitrogen fixation. The process of denitrification is carried out by specialized bacteria that reduce the N in nitrate, extracting the oxygen and using it to oxidize organic material. Most often, the end product of denitrification is molecular nitrogen, and in this case the process has an overall stoichiometry of Reaction (R8.17):

$$NO_3^- + 1\tfrac{1}{4}\text{``CH}_2\text{O''} + H^+ \rightarrow \tfrac{1}{2}(N_2)_g + 1\tfrac{1}{4}(CO_2)_g + 1\tfrac{3}{4}H_2O \quad \text{(R8.17)}$$

However, a small fraction (generally a few percent) of the N involved in denitrification is converted to N_2O instead of N_2, and in these cases the overall stoichiometry is given by Reaction (R8.18):

$$NO_3^- + \text{``CH}_2\text{O''} + H^+ \rightarrow \tfrac{1}{2}(N_2O)_g + (CO_2)_g + 1\tfrac{1}{2}H_2O \quad \text{(R8.18)}$$

Both reactions are exothermic, Reaction (R8.17) more so than (R8.18), and the energy released is used by denitrifiers to support their metabolic functions in the same way that we respirators use the energy from the reaction of oxygen with organic matter. However, denitrification does not yield as much energy as respiration (see Table 3.5), and the process is therefore competitively disadvantageous in aerobic environments. As a result, denitrification is limited to anaerobic environments, where the absence of free oxygen precludes respiration. In fact, denitrification is usually carried out by so-called *facultative anaerobic bacteria*. The word "facultative" is used to denote bacteria that can exist in aerobic environments, where they respire, as well as anaerobic environments, where they depend upon oxidants other than molecular oxygen to drive their metabolic functions.

8.4. THE PREINDUSTRIAL, STEADY-STATE N CYCLE

A schematic illustration of the global, preindustrial, steady-state, cycle of N is illustrated in Figure 8.6. We use an eight-box scheme to represent this cycle, with reservoirs included for atmospheric N_2, N_2O, and NO_y, terrestrial biospheric N and soil N, oceanic biospheric N and inorganic N, and sedimentary N. The values adopted for each of the reservoir amounts and the rates of exchange between reservoirs adopted here are documented in Table 8.3 and 8.4, respectively.

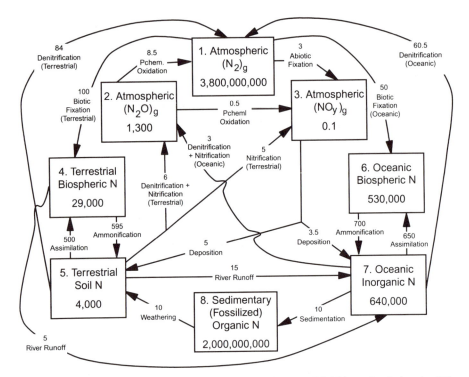

Figure 8.6. The eight-box model for the preindustrial steady-state, global biogeochemical cycle of N. See Tables 8.3 and 8.4 for details.

One of the unique features of the N cycle is the central role played by the atmosphere. For one, the largest N reservoir in the earth system is found in the atmosphere [i.e., atmospheric $(N_2)_g$]. By contrast, recall that in the case of all the other major nutrient elements, the sediments were found to be the largest reservoir. Moreover, the N cycle is the only cycle where we use more than one atmospheric reservoir. In the cycle depicted in Figure 8.6, there are three separate atmospheric reservoirs [i.e., atmospheric $(N_2O)_g$ and $(NO_y)_g$, in addition to $(N_2)_g$]. This is done so that we can see how the various nitrogen oxides, with their very different atmospheric residence times and environmental effects, respond to perturbations to the N cycle. In fact, a more complete treatment of the N cycle would include a fourth atmospheric reservoir: atmospheric $(NH_3)_g$. However, since the vast majority of the N emitted into the atmosphere as ammonia is simply returned back to the lithosphere and ocean via wet and dry deposition in the same -3 oxidation state, the information that can be gleaned from including this reservoir in a global model is fairly limited. Thus, for simplicity, we have elected to omit the ammonia subcycle from our model of the global N cycle. (The task of adding this subcycle into the global cycle is given as a problem at the end of the chapter.)

A final note with regard to the atmospheric portion of our cycle relates to the atmospheric $(NO_y)_g$. Recall that NO_y compounds have a very short atmospheric residence time (of only a few days). As a result, NO_y is not well-mixed in the atmos-

TABLE 8.3
The Preindustrial, Steady-State Global Cycle of N—Reservoir Amounts

Reservoir	Amount (Tg N)	Notes
1. Atmospheric $(N_2)_g$	3.8×10^9	Derived from Equation (3.9) assuming an atmospheric mixing ratio of 0.78. (Note that the reservoir amount reflects the fact that each N_2 molecule contains two N atoms.)
2. Atmospheric $(N_2O)_g$	1.3×10^3	Derived from Equation (3.9) assuming an atmospheric, preindustrial mixing ratio of 275 ppbv.
3. Atmospheric $(NO_y)_g$	0.1	Rough estimate chosen to give a few day atmospheric residence time for NO_y using flux data in Table 8.4.
4. Terrestrial biospheric N	2.9×10^4	Derived from carbon content of living and dead terrestrial biosphere $(2.3 \times 10^6$ Tg C) assuming a N:C ratio of 9:830 on a per atom basis. (See discussion on biosphere in Chapter 3.)
5. Terrestrial (soil) N	4×10^3	Derived from mass of the crust $(2 \times 10^8$ Tg) assuming a N content of 20 μg N per g of crust.
6. Oceanic biospheric N	5.3×10^5	Derived from carbon content of living and dead oceanic biosphere $(3 \times 10^6$ Tg C) assuming a N:C ratio of 16:106 on a per atom basis. (See discussion on biosphere in Chapter 3.)
7. Oceanic inorganic N	6.4×10^5	Derived from average deep ocean nitrate concentration of 35 μM.
8. Sedimentary N	2×10^9	Derived from organic carbon content of sediments $(20 \times 10^9$ Tg) assuming an N:C ratio of 1:10 on a per atom basis. This ratio was obtained by allowing for somewhat more rapid mineralization of N than C in organic matter sinking through the ocean column (i.e., the ratio of 1:10 is smaller than the 16:106 ratio assumed for organic matter in the ocean).

phere, cannot be characterized by a single global, atmospheric concentration, and, therefore, cannot be accurately treated by a global box model like *BOXES*. For this reason, the results we obtain with *BOXES* for the atmospheric $(NO_y)_g$ reservoir should be viewed as only illustrative and not quantitative.

On the basis of the data in Figure 8.7, the **K** matrix for the preindustrial, steady-state N cycle is given by

$$
\mathbf{K} =
\begin{bmatrix}
-4.03 \times 10^{-8} & 6.54 \times 10^{-3} & 0.0 & 0.0 & 2.1 \times 10^{-2} & 0.0 & 9.45 \times 10^{-5} & 0.0 \\
0.0 & -6.92 \times 10^{-3} & 0.0 & 0.0 & 1.5 \times 10^{-3} & 0.0 & 4.69 \times 10^{-6} & 0.0 \\
7.89 \times 10^{-10} & 3.85 \times 10^{-4} & -85 & 0.0 & 1.25 \times 10^{-3} & 0.0 & 0.0 & 0.0 \\
2.63 \times 10^{-8} & 0.0 & 0.0 & -2.07 \times 10^{-2} & 0.125 & 0.0 & 0.0 & 0.0 \\
0.0 & 0.0 & 50 & 2.05 \times 10^{-2} & -0.1525 & 0.0 & 0.0 & 5.0 \times 10^{-9} \\
1.32 \times 10^{-8} & 0.0 & 0.0 & 0.0 & 0.0 & -1.32 \times 10^{-3} & 1.02 \times 10^{-3} & 0.0 \\
0.0 & 0.0 & 35 & 1.72 \times 10^{-4} & 3.75 \times 10^{-3} & 1.32 \times 10^{-3} & -1.13 \times 10^{-3} & 0.0 \\
0.0 & 0.0 & 0.0 & 0.0 & 0.0 & 0.0 & 1.56 \times 10^{-5} & -5.0 \times 10^{-9}
\end{bmatrix}
$$

TABLE 8.4
The Preindustrial, Steady-State Global Cycle of N—Fluxes

Flux	Rate (Tg N year^{-1})	Notes
$F_{1\to3}$		
From atmospheric $(N_2)_g$ To atmospheric $(NO_y)_g$	3	Abiotic fixation rate obtained from estimates of NO production rate by lightning.
$F_{1\to4}$		
From atmospheric $(N_2)_g$ To terrestrial biosphere	100	Terrestrial biotic fixation rate obtained from representative values appearing in literature.
$F_{1\to6}$		
From atmospheric $(N_2)_g$ To oceanic biosphere	50	Ocean biotic fixation rate obtained from representative values appearing in literature.
$F_{2\to1}$		
From atmospheric $(N_2O)_g$ To atmospheric $(N_2)_g$	8.5	N_2O oxidation rate obtained from atmospheric photochemical model calculations appearing in literature.
$F_{2\to3}$		
From atmospheric $(N_2O)_g$ To atmospheric $(NO_y)_g$	0.5	N_2O oxidation rate obtained from atmospheric photochemical model calculations appearing in literature with ~5% of atmospheric N_2O being converted to NO_y.
$F_{3\to5}$		
From atmospheric $(NO_y)_g$ To terrestrial soil N	5	NO_y deposition rate chosen to yield steady state with 80% (30%) of soil (ocean) NO_x source deposited to soil.
$F_{3\to7}$		
From atmospheric $(NO_y)_g$ To oceanic inorganic N	3.5	See above.
$F_{4\to5}$		
From terrestrial biosphere To terrestrial soil	595	Terrestrial ammonification rate derived to balance N uptake of 600 Tg N year^{-1} (derived from net primary productivity and N:C ratio of 9:830) with 5 Tg N year^{-1} lost via runoff.
$F_{4\to7}$		
From terrestrial biosphere To oceanic inorganic N	5	River runoff rate chosen to yield steady state.
$F_{5\to1}$		
From terrestrial soil To atmospheric $(N_2)_g$	84	N_2 emission rate from terrestrial dentrification chosen to yield steady state.
$F_{3\to2}$		
From terrestrial soil To atmospheric $(N_2O)_g$	6	N_2O emission rates from terrestrial dentrification and nitrification chosen to yield steady state.
$F_{5\to4}$		
From terrestrial soil To atmospheric $(NO_y)g$	5	NO_y emission rate from soil adopted from values appearing in literature.
$F_{5\to4}$		
From terrestrial soil To terrestrial biosphere	500	Assimilation rate derived to balance total N uptake of 600 Tg N year^{-1} (see above) with 100 Tg N year^{-1} coming from biotic fixation.
$F_{5\to7}$		
From terrestrial soil To oceanic inorganic N	15	River runoff rate chosen to yield steady state.
$F_{6\to7}$		
From oceanic biosphere To oceanic inorganic N	700	Oceanic ammonification rate derived to balance N uptake of 700 Tg N year^{-1} (derived from oceanic new production and N:C ratio of 16:106).
$F_{7\to1}$		
From oceanic inorganic N To atmospheric $(N_2)_g$	60.5	N_2 emission rate from oceanic dentrification chosen to yield steady state.
$F_{7\to2}$		
From oceanic inorganic N to atmospheric $(N_2O)_g$	3	N_2O emission rates from oceanic dentrification and nitrification chosen to yield steady state.

TABLE 8.4 (Continued)
The Preindustrial, Steady-State Global Cycle of N—Fluxes

$F_{7\to6}$
  From oceanic inorganic N   650   Assimilation rate derived to balance total N uptake of
  To oceanic biosphere   700 Tg N year^{-1} (see above) with 50 Tg N year^{-1}
$F_{7\to8}$ coming from biotic fixation.
  From oceanic inorganic N   10   Sedimentation rate derived from C burial rate of 100 Tg
  To sediments   C year^{-1} assuming a N:C ratio of 1:10 on a per atom
$F_{8\to5}$ basis as described above.
  From sediments   10   Weathering rate chosen to yield steady state.
  To terrestrial soil N

In the next sections, we will use this model to investigate the response of the N cycle to anthropogenic perturbations.

8.5. NUMERICAL EXPERIMENT—EFFECT AND PERSISTENCE OF ANTHROPOGENIC PERTURBATIONS

The activities of our increasingly populous and industrial society perturb the global biogeochemical cycle of N in three ways: (1) Fertilizer production transfers approximately 80 Tg N year^{-1} from the atmospheric $(N_2)_g$ reservoir to the inorganic soil N reservoir, (2) increased cultivation of legumes by farmers artificially enhances the biotic nitrogen fixation rate by about 40 Tg N year^{-1}, and (3) the burning of fossil fuels adds an additional 30 Tg N year^{-1} to the abiotic nitrogen fixation rate. A summation of these perturbations and their affects on the various flux rates in the N biogeochemical cycle are presented in Table 8.5. It should be apparent from the data in Table 8.5 that humankind is causing significant changes in the rates of transfer of N between the atmosphere and the terrestrial lithosphere and biosphere and

TABLE 8.5
The Effects of Anthropogenic Activities on the Fluxes of the Global Biogeochemical Cycle of N

Anthropogenic Activity	Effect
1. Fertilizer production	Transfers 80 Tg N year^{-1} from atmospheric $(N_2)_g$ reservoir to inorganic soil N reservoir: $F_{1\to5} = 80$ Tg N year^{-1}
2. Legume cultivation	Enhances biotic nitrogen fixation in terrestrial systems by 40 Tg N year^{-1}. This increases transfer rate from atmospheric $(N_2)_g$ reservoir to terrestrial biospheric N reservoir from 100 to 140 Tg N year^{-1}: $F_{1\to4} = 140$ Tg N year^{-1}
3. Fossil fuel burning	Enhances abiotic fixation rate by 30 Tg N year^{-1}. This increases transfer rate from atmospheric $(N_2)_g$ reservoir to atmospheric $(NO_y)_g$ reservoir from 3 to 33 Tg N year^{-1}: $F_{1\to3} = 33$ Tg N year^{-1}

in the overall rate of nitrogen fixation. Have these changes engendered significant perturbations to the partitioning of N on a global scale, and, if so, how long will these perturbations persist after a cessation of anthropogenic activities? Let's use *BOXES* to investigate these questions.

Our investigation of the impact of anthropogenic activities upon the N cycle will follow the same basic methodology we adopted in Chapters 6 and 7 for similar studies related to the cycles of C and S, respectively. We adopt a hypothetical scenario that requires carrying out simulations with *BOXES* in two stages. During Stage 1, we impose the anthropogenic perturbations outlined in Table 8.5 upon the preindustrial, steady-state N cycle and allow these perturbations to persist for 130 years. During Stage 2, we remove the anthropogenic perturbations and allow the cycle to relax back to its pre-industrial conditions. Thus, using the same methodology as in Chapters 6 and 7, the Stage 1 simulation adopts initial reservoir amounts from the preindustrial, steady-state cycle (illustrated in Figure 8.7) and an anthropogenically perturbed K matrix given by

$$
\mathbf{K} = \begin{bmatrix}
\mathbf{-7.98 \times 10^{-8}} & 6.54 \times 10^{-3} & 0.0 & 0.0 & 2.1 \times 10^{-2} & 0.0 & 9.45 \times 10^{-5} & 0.0 \\
0.0 & -6.92 \times 10^{-3} & 0.0 & 0.0 & 1.5 \times 10^{-3} & 0.0 & 4.69 \times 10^{-6} & 0.0 \\
\mathbf{8.68 \times 10^{-9}} & 3.85 \times 10^{-4} & -85 & 0.0 & 1.25 \times 10^{-3} & 0.0 & 0.0 & 0.0 \\
\mathbf{3.68 \times 10^{-8}} & 0.0 & 0.0 & -2.07 \times 10^{-2} & 0.125 & 0.0 & 0.0 & 0.0 \\
\mathbf{2.11 \times 10^{-8}} & 0.0 & 50 & 2.05 \times 10^{-2} & -0.1525 & 0.0 & 0.0 & 5.0 \times 10^{-9} \\
\mathbf{1.32 \times 10^{-8}} & 0.0 & 0.0 & 0.0 & 0.0 & -1.32 \times 10^{-3} & 1.02 \times 10^{-3} & 0.0 \\
0.0 & 0.0 & 35 & 1.72 \times 10^{-4} & 3.75 \times 10^{-3} & 1.32 \times 10^{-3} & -1.13 \times 10^{-3} & 0.0 \\
0.0 & 0.0 & 0.0 & 0.0 & 0.0 & 0.0 & 1.56 \times 10^{-5} & -5.0 \times 10^{-9}
\end{bmatrix}
$$

where, as in previous cases, we use a bold font to indicate those matrix elements that have changed from the preindustrial, steady-state cycle. Note that all anthropogenic perturbations involve a change in the rate of N transfer from the atmospheric $(N_2)_g$ reservoir, so all the changes to the **K** matrix appear in column 1. In Stage 2, we use the reservoir contents obtained at $t = 130$ years of the Stage 1 simulation for initial conditions and the original preindustrial, steady-state **K** matrix.

The resulting relative reservoir contents for atmospheric $(NO_y)_g$ and for atmospheric $(N_2O)_g$, the terrestrial biosphere, and inorganic soil N obtained from the combined Stage 1 and Stage 2 simulations are illustrated in Figures 8.7 and 8.8, respectively. (The changes in the other reservoirs are quite small and are not illustrated.) The largest change is obtained for the atmospheric $(NO_y)_g$ reservoir, which increases by more than a factor of four during the Stage 1 simulation. This increase is directly attributable to the sizable increase in the abiotic nitrogen fixa-

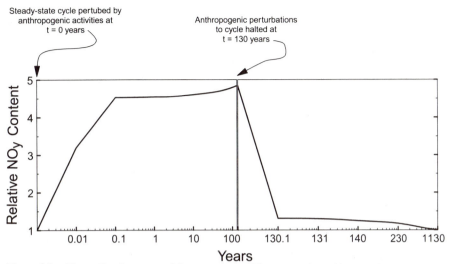

Figure 8.7. The predicted response of the atmospheric $(NO_y)_g$ reservoir to 130 years of anthropogenic perturbations. The large increase and rapid response are characteristic of a reservoir with a short residence time.

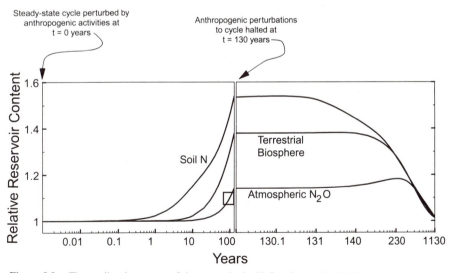

Figure 8.8. The predicted response of the atmospheric $(N_2O)_g$, inorganic soil N, and terrestrial biosphere reservoirs to 130 years of anthropogenic perturbations. Note the more modest increases, relative to those predicted for atmospheric $(NO_y)_g$, but the much longer time required for dissipation of the perturbation. The open box in the figure represents the enhancement in atmospheric N_2O concentrations since the Industrial Revolution inferred from measurements of ambient air and air trapped in ice cores. Does the excellent agreement between the model predicted and observed N_2O enhancement indicate that we have a complete and accurate understanding of the global N cycle?

tion from fossil fuel burning. Since this reservoir has such a short residence time (of only a few days), its response to the anthropogenic perturbations and to the cessation of these perturbations appears to occur almost instantaneously for the time scales adopted in Figure 8.7. However, when viewing these it is important to bear in mind the caveat noted earlier with regard to the simulation of the atmospheric $(NO_y)_g$ reservoir using *BOXES*. Because NO_y has such a short atmospheric residence time, it is not well-mixed within the atmosphere and thus cannot be accurately treated in a global box model like *BOXES*. The results in Figure 8.7 should therefore not be taken too literally. All we can really glean from these simulations is that (i) significant increases in atmospheric NO_y concentrations have probably occurred as a result of anthropogenic emissions (and this is certainly supported by observations of greatly enhanced levels of NO_y in and around the world's urban cores) and (ii) this increase would very quickly dissipate if NO_x emissions from fossil fuel burning were to suddenly cease.

In contrast to the changes in the atmospheric $(NO_y)_g$ reservoir, the perturbations predicted by *BOXES* for atmospheric $(N_2O)_g$, the terrestrial biosphere, and inorganic soil N are more modest (i.e., of the order of 10–50%) but far more persistent, requiring many centuries to dissipate after a return to preindustrial conditions. The slow response is especially noteworthy for atmospheric $(N_2O)_g$, which continues to grow for 200–300 years after the cessation of anthropogenic perturbations as the excess N stored in terrestrial soils continues to drive enhanced rates of denitrification and nitrification.

The persistence of the perturbations to atmospheric $(N_2O)_g$, the terrestrial biosphere, and inorganic soil N makes them far more troubling than the much larger but shorter-lived perturbations predicted for NO_y. If, at some point in the future, society should decide that these perturbations are unacceptable—either because of the deleterious effects of enhanced N_2O (see Table 8.2) or because of unforeseen ecological effects from increased N storage in soil—and acts to cease all activities affecting the N cycle, the world will still continue to face the consequences of these changes for many generations.

8.6. CONCLUSION

Nitrogen occupies a unique position in the earth system. In addition to being one of the major nutrient elements and thus critical to all biospheric processes, the element comprises about 80% of the atmosphere. Moreover, because of the importance of N-containing trace species in the atmosphere, the element's biogeochemical cycle also affects climate, local and regional air quality, the acidity of precipitation, and the stratospheric ozone layer. It is therefore not comforting to find that anthropogenic activities have such a very significant impact on the global cycle and that, because of the time scales involved in the cycle, some of the perturbations to the N cycle from anthropogenic activities will likely persist for centuries after these activities cease.

Given the importance of the global N cycle and the magnitude of the perturbations predicted by *BOXES*, it is relevant to ask ourselves if we really understand everything we need to know about this cycle. One way to answer this question is to see how well predictions from *BOXES* compare to observed quantities. If we go back to Figure 8.9, we see that the predicted change in atmospheric $(N_2O)_g$ after 130 years of anthropogenic perturbations (of about 14%) is in excellent agreement with the observed change in atmospheric N_2O concentrations since the Industrial Revolution (of about 13%). This agreement would appear to suggest that our model of the global N cycle using *BOXES* is accurate and that we have an excellent handle on how this cycle works.

Actually, the situation is not that encouraging, and the agreement between observed and predicted N_2O changes is most likely one of good fortune rather than good science. For instance, if you study Table 8.4 in detail you will see that we forced our preindustrial cycle to be in steady state by specifying a number of critical fluxes, including the N_2O emission rates from denitrification and nitrification. In fact, a careful accounting of the N_2O budget does not yield a balance but a deficit of about 2 Tg N year^{-1} (see Table 8.6). Moreover, although *BOXES* predicts a significant increase in terrestrial biospheric and soil N as a result of anthropogenic emissions, detailed ecological studies have yet to confirm that this is the case. This has in turn given rise to speculation about and the search for a new category of nitrogen—the "missing nitrogen" that, like "missing carbon," has been fixed by anthropogenic activity but has yet to be found within the earth system.[3] All this would seem to imply that there is still much to be learned about the global biogeochemical cycle of N and its response to the very sizable perturbations being wrought by our modern industrial society. Will we find the anthropogenic perturbations to the global N cycle to be more serious or less serious than those predicted by our simple simulations using *BOXES*? Only time and hard work will tell.

TABLE 8.6
The Budget to Atmospheric N_2O: Indications That Something Is Amiss with Our Understanding of the Global N Cycle?

	Flux (Tg N year^{-1})
1. Sum of all known N_2O sources (i.e., emissions from terrestrial and oceanic denitrification and nitrification)	10
2. Sum of all know N_2O sinks (i.e., loss from stratospheric photochemistry)	9
3. Observed rate of increase in atmosphere	3
4. N_2O "deficit" (i.e., sinks + increase − sources)	2

Source: Intergovernmental Panel on Climate Change, *Climate Change 1994: Radiative Forcing of Climate Change,* Cambridge University Press, New York, 1995.

---

[3]The term "missing nitrogen" appears to have been first used in a recent paper by Dr. James Galloway and colleagues (Galloway, J. N., W. H. Schlesinger, H. Levy, II, A. Michaels, and J. L. Schnoor, Nitrogen fixation: Anthropogenic enhancement-environmental response, *Global Biogeochemical Cycles,* **9,** 235–252, 1995.)

SUGGESTED READING

Delwiche, C. C., The nitrogen cycle, in *The Biosphere*, W. H. Freeman, San Francisco, 1970, pp. 69–80.

Dentner, F. J., and P. J. Crutzen, A three-dimensional model of the global ammonia cycle, *Journal of Atmospheric Chemistry*, **19**, 331–369, 1994.

Galloway, J. N., W. H. Schlesinger, H. Levy II, A. Michaels, and J. L. Schnoor, Nitrogen fixation: Anthropogenic enhancement—environmental response, *Global Biogeochemical Cycles*, **9**, 235–252, 1995.

Intergovernmental Panel on Climate Change, *Climate Change 1994: Radiative Forcing of Climate Change*, Cambridge University Press, New York, 1995.

PROBLEM

1. The global rate of volatilization of ammonia from terrestrial soils is estimated at 100 Tg N year^{-1}, and the atmospheric residence time of ammonia is about 10 days. With this information develop a nine-box model for the global N cycle that includes the eight reservoirs treated in Figure 8.7 plus a reservoir for atmospheric ammonia and, using *BOXES*, repeat the numerical experiment investigating the impact of anthropogenic perturbations to the global N cycle. By comparing the rates of deposition of NO_y and NH_3, determine how anthropogenic perturbations affect the acidity of rainfall; specifically determine if these perturbations increase or decrease the acidity of rainfall and by how much as a function of time. Which aspects of the anthropogenic perturbation contribute most to increasing acidity, and which aspects contribute most to decreasing acidity?

Bringing It All Together: The Stability of Atmospheric Oxygen

9

> "As oxygen increases in abundance the growth of consumers increases, but oxygen in excess is poisonous. Too little oxygen and too much are both bad; there is a desirable sufficiency."
>
> J. Lovelock, *The Ages of Gaia*, W. W. Norton, London, 1988.

9.1. INTRODUCTION

In Chapter 1 we began our investigation of global biogeochemical cycles with a discussion of the production and destruction of atmospheric oxygen (O_2) by photosynthesis and respiration. Now in this, our final chapter, we conclude our investigation by returning again to the cycle of atmospheric oxygen. Only this time we will delve more deeply into the cycle. Our purpose: To develop a quantitative model of the processes that determine atmospheric O_2 abundances on time scales ranging from hundreds to millions of years. To accomplish this, we will have to revisit the cycles of each of the major nutrient elements discussed earlier and reconstruct these cycles in a new and more complex manner—one that explicitly couples the individual elemental cycles together into a single, global metabolic system. For this reason, the oxygen cycle represents a fitting topic for the conclusion of our investigation of global biogeochemical cycles.

Of all the nutrient elements, the global biogeochemical cycle of oxygen is probably the most complex and the most intriguing. Oxygen is by far the most abundant element in the earth system, comprising more than 45% of the total mass of the planet. The vast majority of this oxygen is found in its expected -2 oxidation state—primarily in the lithosphere as silicates and metal oxides and, to a lesser extent, in the ocean as water. Fortunately, however, this is not terrestrial oxygen's only oxidation state. As we already know, roughly 20% of the earth's atmosphere is comprised of molecular oxygen (O_2), with an oxidation state of 0. For we respiring earth-

179

lings, who have grown accustomed to the luxury of a plentiful supply of oxygen to breathe, this may not seem to be especially remarkable, but it is. At the very least it is a striking example of the earth's uniqueness because ours is the only planet of the solar system to have more than trace amounts of O_2 in its atmosphere. But the extraordinariness of atmospheric oxygen goes beyond its mere presence to its actual abundance. It seems that the amount of oxygen in the atmosphere just happens to fall within relatively tight constraints for the maintenance of life. If there were very much less oxygen, large animals such as ourselves could not survive. On the other hand, if there were very much more oxygen, widespread fires would become commonplace and likely cause the destruction of much of the terrestrial biosphere. What are the processes that produce and maintain this happy balance between too much and too little atmosphere O_2? Can these processes be perturbed in a way that upsets this balance? And, if so, on what time scales can we expect to see significant fluctuations in the abundance of atmospheric O_2 from these perturbations? These are the questions we will focus on in this chapter. We begin by discussing the biosphere's key role in the oxygen cycle.

9.2. THE OXYGEN CYCLE ON SHORT TIME SCALES: THE BIOSPHERIC CONNECTION

Any discussion of atmospheric O_2 must necessarily include the biosphere, because it is the biosphere that is responsible for the production of atmospheric O_2 via photosynthesis

$$CO_2 + H_2O + h\nu \rightarrow \text{``CH}_2\text{O''} + O_2 \qquad (R1.1)$$

and much of its removal via respiration and decay

$$\text{``CH}_2\text{O''} + O_2 \rightarrow CO_2 + H_2O \qquad (R1.2)$$

These two processes, photosynthesis and respiration, are responsible for cycling more than 99% of the O_2 that passes through the atmosphere each year. It is therefore tempting to try to develop a description of the global oxygen cycle on the basis of these processes alone. Figure 9.1 represents an attempt to do just that. The figure, a schematic of a five-box cycle, captures the essential features of the biosphere's effect on atmospheric O_2. The five reservoirs are as follows: atmospheric O_2, the terrestrial and oceanic biospheres, and atmospheric and oceanic (i.e., dissolved) CO_2. The reservoir amounts and fluxes illustrated in the figure were all derived from data previously discussed; the details of the derivations are summarized in Tables 9.1 and 9.2. In reviewing Figure 9.1 and Tables 9.1 and 9.2, the reader should make note of a few points. First, note that we have chosen to express the reservoir amounts and the flux rates in molar units,[1] instead of mass units. This is done so that we can use stoichiometry to relate the flux of one element to that of another element, without having to worry about different gram-molecular weights.

---

[1] We will express the reservoirs and fluxes in the oxygen cycle in multiples of kTmoles and kTmoles year^{-1}, respectively, where 1 kTmole = 1 kilo-teramole = 1000 teramoles = 10^{15} moles.

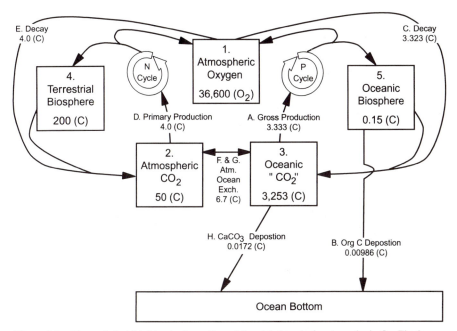

Figure 9.1. The preindustrial, biospheric portion of the global cycle for atmospheric O_2. The boxes represent reservoirs in kTmoles, and the arrows represent fluxes in kTmoles year^{-1}. Couplings to the N and P cycle are indicated schematically by the circle arrows. This portion of the cycle is responsible for more than 99% of the O_2 that cycles through the atmosphere and drives variations in atmospheric O_2 that occur on time scales of hundreds of years or less. (Note: 1 kTmole = 1×10^{15} moles.)

[For example, from Reaction (R1.1) we know that for each mole of CO_2 transferred to the marine biosphere, 1 mole of O_2 is produced. Thus, a value of 3.333 kTmoles year^{-1} of C for F_A, the rate of photosynthesis by the marine biosphere, can be directly converted into an atmospheric oxygen production rate of 3.333 kTmoles year^{-1} of O_2.] Secondly, note that we have not used the $F_{i \to j}$ and $k_{i \to j}$ notation to represent fluxes from one reservoir to another. This notation is only appropriate for

TABLE 9.1
The Preindustrial Global Cycle for Atmospheric O_2: Reservoirs Active on Short Time Scales (i.e., $\tau \sim 100$ years)

Reservoir	Amount (kT moles)	Notes
1. Atmospheric $(O_2)_g$	36,600 (as O_2)	Derived from Equation (3.9) assuming an atmospheric mixing ratio of 0.21.
2. Atmospheric CO_2	50 (as C)	See Chapter 6.
3. Oceanic CO_2	3,253 (as C)	Combines C reservoirs of surface and deep ocean. See Chapter 6.
4. Terrestrial biosphere	200 (as C)	Combines C reservoirs of living and dead terrestrial biosphere. See Chapter 6.
5. Oceanic biosphere	0.15 (as C)	See Chapter 6.

TABLE 9.2

The Preindustrial Global Cycle for Atmospheric O_2: Fluxes Active on Short Time Scales (i.e., $\tau \sim 100$ years)

Flux	Rate (KTmoles year^{-1})	Notes
A. Gross production (ocean)	3.333 (as C)	See Chapter 6.
B. Organic C deposition	0.00986 (as C)	Total deposition of organic C at the ocean bottom assumed to be the sum of (i) The burial rate of organic C of 0.00833 kTmoles C year^{-1} and (ii) (15/8) times the rate of burial of FeS$_2$ of 0.000817 kTmoles S year^{-1} (i.e., $F_B = F_J + (15/8)F_I$). See Chapters 6 and 7 and discussion in Section 9.3.
C. Decay of oceanic biosphere	3.323 (as C)	Decay rate specified to be the difference between (i) The gross ocean production of 3.333 kTmoles C year^{-1} and (ii) The organic C deposition rate of 0.00986 kTmoles C year^{-1} (i.e., $F_C = F_A - F_B$).
D. Primary production (terrestrial)	4.0 (as C)	See Chapter 6.
E. Decay, terrestrial biosphere	4.0 (as C)	See Chapter 6.
F. Atmospheric CO_2 to ocean transfer	6.6667 (as C)	See Chapter 6.
G. Oceanic CO_2 to atmosphere transfer	6.6658 (as C)	CO_2 flux from ocean to atmosphere specified to be the difference between (i) The CO_2 flux from atmosphere to ocean of 6.6667 kTmoles C year^{-1} and (ii) The rate of burial of organic C of 0.0083 kTmoles C year^{-1} (i.e., $F_G = F_F - F_J$). See Chapter 6 and discussion in Section 9.3.
H. CaCO$_3$ deposition	0.0172 (as C)	Total deposition of CaCO$_3$ assumed to be the sum of (i) the burial rate of CaCO$_3$ of 0.01667 kTmoles C year^{-1} and (ii) the burial rate of CaSO$_4$ of 0.00053 kTmoles S year^{-1} (i.e., $F_H = F_K + F_L$). See Chapters 6 and 7 and discussion in Section 9.3.

linear systems, in which all the fluxes are linearly dependent upon the size of the reservoir from which they emanate. As we shall quickly see, many of the fluxes in the oxygen cycle require more complex functional relationships, and thus the simple linear notation we have been using is no longer useful.

Inspection of Figure 9.1 reveals a relatively simple picture for the oxygen cycle at this level of detail. O_2 is produced whenever C is transferred from the atmospheric or oceanic CO_2 reservoirs to the terrestrial or oceanic biospheres, and O_2 is removed whenever biospheric C is returned to one of the two CO_2 reservoirs. To take into account the distinction between photosynthesis in the ocean and on land, we distinguish between CO_2 in the atmosphere and dissolved CO_2 in the ocean (i.e., Reservoirs 2 and 3) and we include fluxes that move C between these two reservoirs. The reader will note that we have simplified our treatment of the C portion of the cycle by combining the surface and deep ocean CO_2 reservoirs into a single "oceanic 'CO_2' reservoir" (i.e., Reservoir 3) and combining the living and dead terrestrial biosphere into a single "terrestrial biosphere" (i.e., Reservoir 4). Because our terrestrial biosphere includes both living and dead organic material, we need only consider net primary production of the terrestrial biosphere, as that is the only process that leads to long-term storage of organic material on the continents.

The five-box model illustrated in Figure 9.1 can explain much of the variability in atmospheric O_2 that occurs on timescales of hundreds of years or less. However, it has some obvious limitations. In the first place, the cycle is not closed and it is therefore not in steady state. There is a small, but nevertheless steady, leakage of material to the ocean bottom as organic C and calcium carbonate ($CaCO_3$). If we were to attempt to model this cycle, we would find that eventually all the C would be drained from the atmosphere and ocean to the ocean bottom. At the same time, we would find a slight increase in atmospheric O_2 as a result of the small imbalance between oceanic gross production and the decay of the oceanic biosphere.

A second and related problem has to do with an apparent stoichiometric imbalance in the C and O_2 reservoirs. If all the O_2 in the atmosphere originated from photosynthesis, then we might expect the number of moles of O_2 in the atmosphere to be balanced by the number of moles of organic C (or related reduced material) in the earth system. In our simple model, however, there are orders of magnitude more O_2 than organic C in the marine and terrestrial biosphere. The remedy for both of these limitations can be found in the rock cycle, the subject of the next section.

9.3. THE OXYGEN CYCLE ON LONG TIME SCALES: THE ROCK CYCLE CONNECTION

Figure 9.2 illustrates the coupling of the short-time-scale, biospheric portion of the oxygen cycle to its longer-term components contained within the rock cycle.[2] In this model the oxygen cycle's connection to the rock cycle is accomplished through two pathways, both of which involve interactions with the global S cycle. The first pathway arises from the deposition of organic C on the ocean bottom (i.e., F_B). Once deposited on the ocean floor, the organic C can be buried directly, leading to the formation of organic C sediments (i.e., F_J), or it can be processed by colorless bacteria in the presence of sulfate and ferric iron (e.g., hematite) to form pyrite (FeS_2) sediments (i.e., F_I) via[3]

$$8SO_4^{2-} + 2(Fe_2O_3)_S + 15\text{“CH}_2\text{O”} + 16H^+$$
$$\rightarrow 4(FeS_2)_S + 15CO_2 + 23H_2O \qquad (R7.4)$$

Note that both organic C and FeS_2 sediments represent material in a reduced oxidation state. Because this reduced oxidation state was originally generated by the production of the "CH_2O" during photosynthesis, the burial of these sediments represents a net source of O_2 to the atmosphere. Specifically, 1 mole of O_2 is left in the atmosphere for each mole of organic C buried and 15/8 moles of O_2 are left in

---

[2]The model we adopt for coupling the oxygen cycle to the ocean sediments and the rock cycle via the S cycle closely follows the model of Garrels et al. described in a paper published in the *American Scientist* in 1976. An interesting alternate approach involving P can be found in a paper by Van Cappellen and Ingall published in *Science* in 1996.

[3]The reader may find it helpful at this point to briefly review the discussion in Chapter 7 on the formation of pyrite and gypsum sediments and their potential impact on atmospheric O_2.

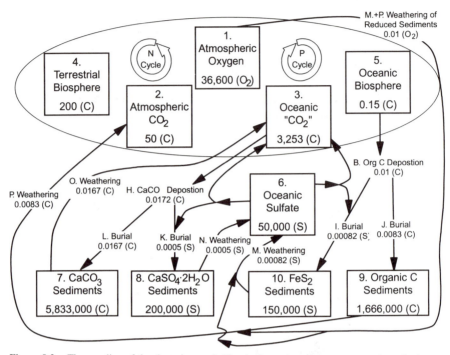

Figure 9.2. The coupling of the short-time-scale biospheric portion of the oxygen cycle to the longer timescale rock cycle. As in Figure 9.1, the boxes represent reservoirs in kTmoles and the arrows represent fluxes in kTmoles year^{-1}. The biospheric portion of the cycle, illustrated in more detail in Figure 9.1, consists of Reservoirs 1–5 and appears within the oval. The remaining portion of the cycle is responsible for less than 1% of the O_2 that cycles annually through the atmosphere, but controls variations in atmospheric O_2 that occur on time scales of millions of years. (Note: 1 kTmole $= 1 \times 10^{15}$ moles.)

the atmosphere for each mole of S buried as FeS_2. Moreover, this O_2 must remain in the atmosphere until these sediments are brought to the earth's surface and weathered via the standard respiration and decay reaction in the case of the organic C sediment (i.e., F_P), or via

$$4FeS_2 + 8H_2O + 15O_2 \rightarrow 2Fe_2O_3 + 8SO_4^{2-} + 16H^+ \qquad (R7.5)$$

in the case of the FeS_2 sediments (i.e., F_M).

The other pathway for connecting the oxygen cycle to the rock cycle is through the deposition of $CaCO_3$ on the ocean bottom (i.e., F_H). Similar to the deposition of organic C, the $CaCO_3$ deposited on the ocean floor can be buried directly—in this case forming calcite sediments (i.e., F_L)—or it can interact with sulfate to form gypsum sediments (i.e., F_K) via

$$(CaCO_3)_S + H^+ + 2H_2O + SO_4^{2-} \rightarrow (CaSO_4 \cdot 2H_2O)_S + HCO_3^- \qquad (R7.7)$$

The C and S sediments formed by these processes remain in the lithosphere until they are brought to the earth's surface and weathered, thereby returning the dissolved CO_2 and sulfate to the ocean and closing the C and S cycles. Note that neither the formation or weathering of the $CaCO_3$ and $CaSO_4 \cdot 2H_2O$ sediments has any direct effect on atmospheric O_2 (see Tables 9.3 and 9.4). Nevertheless, because these

TABLE 9.3
The Preindustrial Global Cycle for Atmospheric O_2: Reservoirs Active on Long Time Scales (i.e., $\tau \geq 10^6$ years)

Reservoir	Amount (kTmoles)[a]
6. Oceanic sulfate (SO_4^{ocean})	50,000 (as S)
7. Calcite sediments ($CaCO_3$)$_s$	5,833,000 (as C)
8. Gypsum sediments ($CaSO_4 \cdot 2H_2O$)$_s$	200,000 (as S)
9. Organic C sediments (Org C)$_s$	1,666,000 (as C)
10. Pyrite sediments (FeS_2)$_s$	150,000 (as S)

[a]Reservoir amounts were taken from data discussed in Chapters 6 and 7.

processes affect the abundances of oceanic CO_2 and sulfate, which do have an impact on atmospheric O_2, they need to be considered in our model of the global oxygen cycle.

9.4. BUILDING A MATHEMATICAL MODEL FOR THE OXYGEN CYCLE

Having identified the components of the global oxygen cycle, we now turn to the task of constructing a mathematical model of the cycle. The first step in this task is to write down the differential equations describing the time rate of change of each of the 10 reservoirs in our cycle.

TABLE 9.4
The Preindustrial Global Cycle for Atmospheric O_2: Fluxes Active on Long Time Scales (i.e., $\tau \geq 10^6$ years)

Flux	Rate[a] (kTmoles year^{-1})
I. Burial of (FeS_2)$_s$	0.00082 (as S)
J. Burial of (Org C)$_s$	0.00833 (as C)
K. Burial of ($CaSO_4 \cdot 2H_2O$)$_s$	0.00053 (as S)
L. Burial of ($CaCO_3$)$_s$	0.01667 (as C)
M. Weathering of (FeS_2)$_s$	0.00082 (as S)
N. Weathering of ($CaSO_4 \cdot 2H_2O$)$_s$	0.00053 (as S)
O. Weathering of ($CaCO_3$)$_s$	0.01667 (as C)
P. Weathering of (Org C)$_s$	0.00833 (as C)

[a]Flux rates were taken from data discussed in Chapters 6 and 7.

9.4.1. The Differential Equations

For the most part, the differential equations can be derived by simply adding and subtracting all the fluxes in Figures 9.1 and 9.2 that flow into and out of each of the boxes. For example, for the atmospheric O_2 reservoir the sources are photosynthesis (i.e., F_A and F_D) and the sinks are biospheric decay (i.e., F_C and F_E) and weathering of reduced sediments (i.e., F_M and F_P). Thus, we obtain

$$\frac{dC(O_2^{\text{atm}})}{dt} = (F_A + F_D) - F_C - F_E - F_M - F_P \tag{9.1}$$

Similarly for the atmospheric CO_2 reservoir:

$$\frac{dC(CO_2^{\text{atm}})}{dt} = (F_E + F_G + F_P) - F_D - F_F \tag{9.2}$$

Some care is required for the oceanic CO_2 reservoir because it is produced via F_I. Because F_I, the rate of pyrite burial via Reaction (R7.4), is expressed in moles of S, we must multiply this flux by a factor of (15/8) to get the correct rate of release of oceanic CO_2. On the other hand, because the stoichiometry between gypsum burial and CO_2 release via Reaction (R7.7) is 1:1, no multiplicative factor is needed for F_K. Hence,

$$\frac{dC(CO_2^{\text{ocean}})}{dt} = \left(F_C + F_F + \left(\frac{15}{8}\right)F_I + F_K + F_O\right) - F_A - F_H \tag{9.3}$$

Following a similar procedure, the differential equations for the remaining reservoirs are easily obtained:

$$\frac{dC(\text{Terr. Biosph.})}{dt} = F_D - F_E \tag{9.4}$$

$$\frac{dC(\text{Oc. Biosph.})}{dt} = F_A - F_B - F_C \tag{9.5}$$

$$\frac{dC(SO_4^{\text{ocean}})}{dt} = (F_M + F_N) - F_I - F_K \tag{9.6}$$

$$\frac{dC(CaCO_3)}{dt} = F_L - F_O \tag{9.7}$$

$$\frac{dC(CaSO_4 \cdot 2H_2O)}{dt} = F_K - F_N \tag{9.8}$$

$$\frac{dC(\text{Org. C})}{dt} = F_J - F_P \tag{9.9}$$

$$\frac{dC(FeS_2)}{dt} = F_I - F_M \tag{9.10}$$

9.4.2. Deriving Mathematical Expressions for the Fluxes

Now, having written down the differential equations for the 10 reservoirs in our cycle, we turn to the second step in the construction of our model: namely, deciding how to mathematically express each of the flux terms that appear in Equations (9.1)–(9.10). Without question, this is the more difficult and challenging step in our model development, because there is no single, correct way to express the flux terms. We are only limited by our understanding of the fundamental biogeochemical processes, our willingness to experiment, and our ability to deal with complexity. The expressions derived below comprise but one of many possible paths we could follow.

9.4.2.1. Gross Oceanic Production

Complications immediately arise when we undertake to represent the rates of photosynthesis. Recall that in each of the global cycles we developed earlier, we assumed that the rate of photosynthesis was proportional to the amount of the specific element we were treating at the time. Now, as we allow for couplings between the individual cycles, we are going to have to be a little more rigorous. To decide what to do about the rate of photosynthesis in the ocean, let us return to an argument that was advanced back in Chapter 5 on the global cycle of P. In that chapter, we had remarked on the fact that the N:P ratio in the ocean closely mimics the N:P ratio of ocean biomass. We speculated that this was due to the action of blue-green algae, who fixed just enough nitrogen to balance the availability of ocean phosphate and thereby maintain a sufficient supply of nitrate for the marine biosphere. If true, this implies that photosynthesis in the ocean is ultimately limited by phosphate; and this, in turn, means that we must include a coupling of our global oxygen cycle to the global cycle of P. To accomplish this, we will assume that F_A, the rate of ocean photosynthesis, is linearly dependent upon the amount of phosphate in the ocean; that is,

$$F_A = r_A C(PO_4^{ocean}) \text{ [kTmoles C year}^{-1}] \tag{9.11}$$

where r_A is a constant and $C(PO_4^{ocean})$ is the total abundance of phosphate in the ocean in kTmoles P.

We now have to specify the constant r_A. To do this we demand that $F_A = 3.333$ kTmoles C year^{-1} (see Table 9.1) for preindustrial conditions. Given a preindustrial ocean phosphate abundance of 3 kTmoles P (see Chapter 5), we can easily solve for r_A by substitution into Equation (9.11) and we obtain

$$r_A = 1.111 \quad (\text{year}^{-1}) \tag{9.12}$$

9.4.2.2. Organic C Deposition

For each mole of organic C deposited on the ocean bottom and buried as either reduced C or reduced S, 1 mole of O_2 is left behind in the atmosphere. Moreover, this mole of O_2 remains in the atmosphere for a period of the order of millions of years awaiting the return of the reduced sediment to the earth's surface. Thus, the expression we choose for representing the rate of organic C deposition will play a key role in determining how our model will behave. Observations indicate that the availability of O_2 is critical in determining the fraction of the C fixed by phytoplankton that avoids oxidation and is deposited on the ocean floor. In a world with an unlimited supply of O_2, no C would be deposited, and in a world with no O_2, all the C would be deposited. After some experimentation, we have found that this relationship can best be represented in our model using an exponential function; that is,

$$F_B = F_A \exp[-r_B C(O_2{}^{atm})] \qquad (\text{kTmoles C year}^{-1}) \qquad (9.13)$$

where r_B is a constant. Now, demanding that $F_B = 0.00986$ kTmoles yr^{-1} and assuming that for preindustrial conditions $F_A = 3.333$ kTmoles year^{-1} and $C(O_2{}^{atm}) = 36,600$ kTmoles, we can solve for r_B from Equation (9.13):

$$r_B = 1.591 \times 10^{-4} \qquad (\text{kTmoles}^{-1}) \qquad (9.14)$$

thereby giving us a complete expression for the rate of organic C deposition.

9.4.2.3. Decay of Marine Biosphere

In order to ensure a steady-state cycle for preindustrial conditions, we demand that the decay of the marine biosphere equal the difference between primary production and the deposition of organic C. Thus,

$$F_C = F_A - F_B = F_A (1 - \exp[-r_B C(O_2{}^{atm})]) \qquad (9.15)$$

9.4.2.4. Net Primary Production of Terrestrial Biosphere

Although we assumed that phosphate limited ocean photosynthesis, this would not be appropriate for terrestrial photosynthesis because most of the large, unmanaged terrestrial ecosystems appear to be nitrate limited. We will therefore assume that F_D, the rate of terrestrial primary production, varies linearly with the soil nitrate content, $C(NO_3{}^{soil})$. (Note that, in so doing, we effectively couple our oxygen cycle to the last remaining major nutrient element cycle, namely that of N.). In order to account for the CO_2-fertilization effect on terrestrial biomass, we will also include a weak dependence on atmospheric CO_2 using the β factor discussed in Chapter 6. Thus, our expression for F_D takes the following form:

$$F_D = r_D C(NO_3{}^{soil}) \left[(1 - \beta) + \beta \left(\frac{C(CO_2{}^{atm})}{50} \right) \right] \qquad (\text{kTmoles C year}^{-1}) \qquad (9.16)$$

where r_D is a constant and β defines the relative change in the terrestrial production rate for a doubling in $C(CO_2{}^{atm})$. In our Standard Model calculations we will use $\beta = 0.3$ as we did in our simulations of the C cycle using *BOXES* in Chapter 6. If we now demand that $F_D = 4$ kTmoles C year^{-1} for preindustrial conditions [i.e., $C(NO_3{}^{soil}) = 0.3$ kTmoles N and $C(CO_2{}^{atm}) = 50$ kTmoles C], we find that

$$r_D = 13.333 \qquad (\text{year}^{-1}) \qquad (9.17)$$

9.4.2.5. Decay of Terrestrial Biosphere

To ensure a steady state, we simply demand that

$$F_E = F_D \qquad (9.18)$$

9.4.2.6. Transfer of Atmospheric CO_2 to Ocean

The simplest representation for the flux of atmospheric CO_2 to the ocean would be one that varies linearly with $C(CO_2^{atm})$. However, we know from our discussion of the Revelle factor in Chapter 6 that this flux does not vary linearly with atmospheric CO_2 because of the buffering effect of oceanic carbonate and boric acid. In order to account for this buffering effect, we must incorporate the Revelle factor into our expression for F_F:

$$F_F = r_F \left[\left(1 - \frac{1}{\epsilon} \right) + \frac{1}{\epsilon} \frac{C(CO_2^{atm})}{50} \right] \qquad \text{(kTmoles C year}^{-1}) \qquad (9.19)$$

where r_F is a constant and ϵ is the Revelle factor ($= 10$ in our Standard Model). In order to have $F_F = 6.667$ kTmoles year^{-1} when $C(CO_2^{atm}) = 50$ kTmoles C, we find that

$$r_F = 6.6667 \qquad \text{(kTmoles year}^{-1}) \qquad (9.20)$$

9.4.2.7. Transfer of Oceanic CO_2 to Atmosphere

Having already accounted for the Revelle factor in our expression for F_F, it suffices to assume a linear dependence of the reverse process on oceanic CO_2 ($C(CO_2^{ocean})$); that is,

$$F_G = r_G C(CO_2^{ocean}) \qquad (9.21)$$

In order to obtain a steady state for preindustrial conditions, we demand that

$$F_G = F_F - F_B - F_H \qquad (9.22)$$

and thus

$$r_G = 0.00204 \qquad \text{(year}^{-1}) \qquad (9.23)$$

9.4.2.8. Deposition of $CaCO_3$

The deposition of $CaCO_3$ is assumed to vary linearly with $C(CO_2^{ocean})$. Under this assumption and the requirement that the deposition rate equal 0.0172 kTmoles year^{-1} for $C(CO_2^{ocean}) = 3253$ kTmoles C, we obtain

$$F_H = 5.287 \times 10^{-6} \, C(CO_2^{ocean}) \qquad \text{(kTmoles C year}^{-1}) \qquad (9.24)$$

9.4.2.9. Pyrite Burial

The burial rate of pyrite is assumed to vary linearly with both the deposition rate of organic C and the oceanic sulfate abundance, $C(SO_4^{ocean})$; that is,

$$F_I = r_I \left[\frac{8 \text{ moles S}}{15 \text{ moles C}} \right] F_B C(SO_4^{ocean}) \qquad \text{(kTmoles S year}^{-1}) \qquad (9.25)$$

In order to have $F_I = 0.000817$ kTmoles S year^{-1}, for $F_B = 0.00986$ kTmoles C year^{-1} and $C(SO_4^{ocean}) = 50{,}000$ kTmoles S, it follows that

$$r_I = 3.103 \times 10^{-6} \qquad \text{(kTmoles}^{-1}) \qquad (9.26)$$

9.4.2.10. Burial of Organic C

The rate of burial of organic C is specified to ensure conversation of mass at steady state and thus

$$F_J = F_B - \left[\frac{15 \text{ moles C}}{8 \text{ moles S}} \right] F_I = F_B[1 - r_I C(SO_4^{ocean})]$$

$$(\text{kTmoles C year}^{-1}) \qquad (9.27)$$

9.4.2.11. Gypsum and Calcite Burial

The expressions for burial of gypsum and calcite are derived using the same procedure as that for the burial of pyrite and organic C. In this case, gypsum burial is assumed to vary linearly with oceanic sulfate and the total deposition of $CaCO_3$. The rate of calcite burial is assumed to be the difference between $CaCO_3$ deposition and gypsum burial. Thus,

$$F_K = r_K F_H C(SO_4^{ocean}) \qquad (\text{kTmoles S year}^{-1}) \qquad (9.28)$$

and

$$F_L = F_H - F_K = F_H [1 - r_K C(SO_4^{ocean})] \qquad (\text{kTmoles C year}^{-1}) \qquad (9.29)$$

Demanding that F_K and F_L attain their preindustrial values for preindustrial conditions, we find that

$$r_K = 6.187 \times 10^{-7} \qquad (\text{kTmoles}^{-1}) \qquad (9.30)$$

9.4.2.12. Weathering Rates

The four weathering rates for the pyrite, organic C, gypsum, and calcite sediments are all assumed to vary linearly with their respective abundances. Thus,

$$F_M = 5.440 \times 10^{-9} \, C(FeS_2) \qquad (\text{kTmoles S year}^{-1}) \qquad (9.31)$$

$$F_N = 2.665 \times 10^{-9} \, C(CaSO_4 \cdot 2H_2O) \qquad (\text{kTmoles S year}^{-1}) \qquad (9.32)$$

$$F_O = 2.863 \times 10^{-9} \, C(CaCO_3) \qquad (\text{kTmoles C year}^{-1}) \qquad (9.33)$$

and

$$F_P = 5.018 \times 10^{-9} \, C(Org \, C) \qquad (\text{kTmoles C year}^{-1}) \qquad (9.34)$$

9.4.3. Solving the Equations

Substitution of Equations (9.11)–(9.34) into Equations (9.1)–(9.10) yields a closed system of 10 coupled, first-order differential equations that simulate the time rate of variation in the abundances of the 10 reservoirs of our global oxygen cycle. In order to actually solve for these abundances as a function of time, however, we must integrate this system of differential equations. Unfortunately, because many of the flux terms defined above do not follow the simple linear formulation initially introduced in Chapter 4, we cannot use *BOXES* to do the integration for us. Fortunately, there are many other numerical techniques available to us that can be used to solve our problem. In the subsequent sections, we describe three numerical experiments we have carried out with our model using a simple numerical scheme we developed on our personal computer.[4]

9.5. NUMERICAL EXPERIMENT #1: ENHANCED GYPSUM BURIAL DURING THE PERMIAN REVISITED

In Chapter 7, we carried out a numerical experiment designed to simulate an approximately 50-million-year period during the Permian that was characterized by enhanced burial rates of gypsum and reduced burial rates of pyrite. Using *BOXES* to simulate our uncoupled and linear model for the global S cycle, we calculated that this change in S burial rates over a 50-million-year period would result in the shift of about 20,000 kTmoles of S from the pyrite sediment reservoir to the gypsum sediment reservoir. Because the burial of 1 mole of pyrite represents a net source of 15/8 moles of atmospheric O_2, a shift of this amount of S from the reduced S sediments to the oxidized sediments would effectively remove almost 40,000 kTmoles of O_2 from the atmosphere. Thus the result we obtained in Chapter 7 suggested that, in the absence of any compensating effects, the change in S burial rates during the Permian should have virtually wiped out the atmosphere's entire reservoir of O_2, which is, after all, only 36,600 kTmoles. However, the fossil record indicates that this clearly did not happen. Let's repeat this experiment using our new, coupled model for the global oxygen cycle, to see if we are able to come up with a more plausible explanation for what happened to atmospheric O_2 during the Permian.

9.5.1. Setting Up the Experiment

As detailed in Chapter 7, our experiment involves maintaining the same total rate of S burial (of 0.00135 kTmoles year^{-1}) while shifting the relative rates from their present ratio of 1 mole of pyrite for each 1.5 moles of gypsum to a new, perturbed ratio of 1 mole of pyrite for each 3 moles of gypsum. To accomplish this using our present system of equations, we need to (i) reduce r_I from its standard value of 3.103×10^{-6} kTmoles^{-1} to 1.283×10^{-6} kTmoles^{-1}, (ii) increase r_K from 6.187×10^{-7} to 1.174×10^{-6} kTmoles^{-1}, (iii) initialize the model with the reservoir amounts listed in Tables 9.1 and 9.3, and (iv) integrate the system of equations for 50 million years of simulated time.

---

[4]Rather than supply the reader with the numerical integrator we used (as we have done for the previous cycles), the task of writing an algorithm to integrate the system of equations derived for the oxygen cycle is left as an exercise. J. C. G. Walker's *Numerical Methods in the Earth Sciences* (Cambridge University Press, 1993) provides a number of useful routines that can be used in this regard and thus provides a useful starting point for the less advanced reader. We have found that essentially any integrating scheme will suffice for our system of equations, provided that one chooses small enough time steps to minimize numerical error. The general rule does apply here: The cruder the numerical integration scheme, the smaller the integrating step required.

9.5.2. Results of Numerical Experiment #1

The calculated cumulative changes in the sizes of the four sediment reservoirs and the atmospheric O_2 reservoir during the 50-million-year simulation are illustrated in Figure 9.3. Just as we found when we carried out this numerical experiment using *BOXES* in Chapter 7, changing the relative rates of gypsum and pyrite burial results in a transfer of roughly 20,000 kTmoles of S from the reduced FeS_2 sediment to the oxidized $CaSO_4 \cdot 2H_2O$ sediment by the end of the 50-million-year period (compare Figure 7.8 with Figure 9.3). However, in spite of the sizable change in the S sediment reservoirs, our present model predicts that atmospheric O_2 remains essentially unchanged throughout the period. Given a simple linear picture of the interaction between the cycles of S and O, this result may seem to be implausible. If the burial of FeS_2 sediment represents a source of atmospheric O_2 and the burial of $CaSO_4 \cdot 2H_2O$ sediments does not, how can such a large shift in the burial of these two sediments have no effect on atmospheric O_2? The answer lies in delving deeper into the elemental interactions that comprise the oxygen cycle. Let's begin by examining what happens to the C sediments in our experiment.

The results illustrated in Figure 9.3 indicate a significant shift in the sizes of the

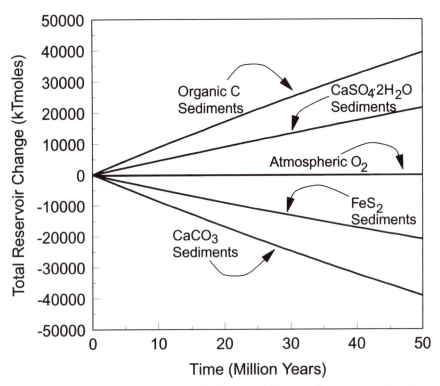

Figure 9.3. Cumulative calculated change in the sizes of the reservoirs of organic C sediments, $CaSO_4 \cdot 2H_2O$ sediments, atmospheric O_2, FeS_2 sediments, and $CaCO_3$ sediments as a function of time when the relative rates of pyrite and gypsum burial are assumed to occur at a ratio of 1:3. We find that the decrease in the burial of reduced S is compensated by a stoichiometrically equivalent increase in the burial of reduced C. The final upshot: no significant change in atmospheric O_2.

C-containing sediments as well as those containing S. In the case of the C sediments, however, we find an increase in the burial of the reduced sediment (i.e., organic C) and a decrease in the burial of the oxidized sediment (i.e., $CaCO_3$). This occurs, of course, because of the exchange of C by S in Reactions (R7.4) and (R7.7) that lead to the formation of the pyrite and gypsum sediments. A reduction in the total rate of FeS_2 burial by 20,000 kTmoles of S leaves 15/8 × 20,000 (about 40,000) kTmoles of organic C on the ocean bottom to be buried as reduced C sediments. The increased burial of organic C sediments results in an equivalent decrease in $CaCO_3$ sediment burial to maintain a C balance in the ocean/atmosphere system. And, although a shift in about 20,000 kTmoles of S from pyrite to gypsum sediments would tend to remove about 40,000 kTmoles of O_2 from the atmosphere, a shift in 40,000 kTmoles of C from calcite to organic C sediment would tend to add the same amount of O_2 to the atmosphere. The result is no net change in the abundance of atmospheric O_2.

Thus our numerical experiment reveals the action of a powerful feedback mechanism in the oxygen cycle—one that compensates for changing rates of burial of one kind of reduced sediment with an equivalent but opposite change in the burial rate of the other reduced sediment. Despite a seemingly overwhelming, long-term perturbation to the system, atmospheric O_2 is stabilized and remains unchanged. Does this mean that the abundance of atmospheric O_2 is impervious to any and all perturbations? In the next section we will attempt to find out by undertaking an experiment that considers the ultimate perturbation: doomsday!

9.6. NUMERICAL EXPERIMENT #2: THE DOOMSDAY SCENARIO

Since all oxygen in the atmosphere originates from photosynthesis, the surest way to deplete the atmospheric O_2 reservoir would be to kill all the world's green plants. But how rapidly would atmospheric O_2 be depleted if such an event were to occur? In this section we will explore this question by simulating this so-called "doomsday scenario" using our model for the oxygen cycle.

We set up this experiment by stopping all oceanic and biospheric production, and this is easily accomplished by setting the flux coefficients for these two processes to zero; that is

$$r_A = r_D = 0 \qquad (9.35)$$

We then set the initial reservoir amounts to the values specified in Tables 9.2 and 9.4 and then integrate the system of equations.

The resulting variations in the biospheric, atmospheric O_2, and atmospheric and oceanic CO_2 reservoirs over the first 150 years of the simulation and over a period of 4 million years are illustrated in Figures 9.4 and 9.5, respectively. Not surprisingly, we find that the halting of all photosynthesis leads to a fairly rapid decay of the biosphere. The disappearance of the oceanic biosphere occurs within the first year (so short a period that its decay cannot be plotted in Figure 9.4), and the disappearance of the terrestrial biosphere occurs in about 150 years.

However, because the amount of C contained within the biosphere is so small

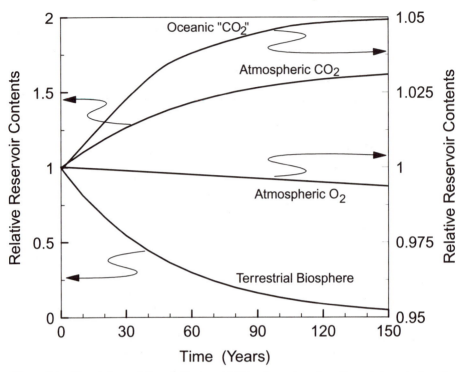

Figure 9.4. The relative variations in the terrestrial biosphere, atmosphere O_2, and atmospheric and oceanic CO_2 reservoirs during the first 150 years after halting all photosynthesis. Note that the oceanic biosphere completely decays within the first year and could not be illustrated here. Although the model predicts the almost total demise of the biosphere within 150 years, less than 1% of the atmospheric O_2 reservoir is lost during this period.

compared to the reservoir of atmospheric O_2, the decay of the biosphere consumes less than 1% of the atmosphere's O_2. As illustrated in Figure 9.5, atmospheric O_2 continues to decline via weathering reactions with pyrite and organic C sediments after the demise of the biosphere, but at a very slow rate. Because of the long time scale associated with the rock cycle, it takes almost 4 million years for the atmosphere's O_2 reservoir to be completely depleted.

This is a truly remarkable result. By sequestering its reduced material deep in the lithosphere, the earth system has given enormous stability to the atmospheric O_2 reservoir. Although catastrophe can lead to the complete loss of atmospheric O_2, it takes millions of years for the process to go to completion. It is almost as if the earth has provided itself and its inhabitants with a 4-million-year "grace period" for recovery from cataclysmic events.

9.7. NUMERICAL EXPERIMENT #3: USING THE OXYGEN CYCLE TO FIND THE "MISSING CARBON"

In the first two numerical experiments we considered fairly remote and esoteric scenarios—enhanced gypsum burial during the Permian and the complete and total

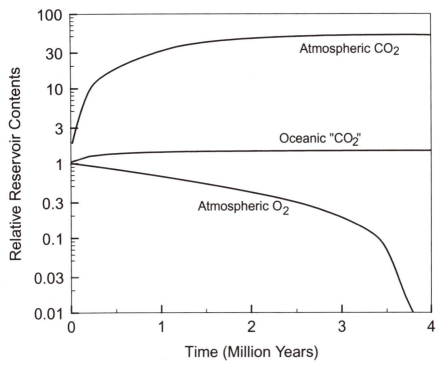

Figure 9.5. The relative variations in the atmospheric O_2 and atmospheric and oceanic CO_2 reservoirs during a 4-million-year period following the destruction of all green plants and the halting of all photosynthesis. Following the demise of the biosphere during the first 150 years (see Figure 9.4), atmospheric O_2 decays slowly via weathering reactions with reduced sediments brought to the earth's surface by the rock cycle. The complete loss of atmospheric O_2 is estimated to take almost 4 million years.

destruction of the biosphere. In this section we will illustrate how an understanding of the global oxygen cycle can aid in addressing a very real and contemporary environmental problem, namely, identifying the fate of the C added to the atmosphere each year by fossil-fuel burning and biomass burning.

The concept of the "missing carbon" was introduced in Chapter 6. Recall that since the Industrial Revolution about 350 Gtons C (or 29 kTmoles C) have been emitted into the atmosphere by anthropogenic activities. Of this total, only about 44% remains in the atmosphere. The rest (the so-called *nonairborne fraction*) has apparently been absorbed by either the ocean or the terrestrial biosphere. There is no problem with roughly half of this nonairborne fraction; marine geochemists can account for the uptake by the ocean of about 100 Gtons C (or 8 kTmoles C) of the excess CO_2 over the past 100 years. The missing C refers to the remaining 100 Gtons C. Where is it? Some believe it has been absorbed by the terrestrial biosphere. (Recall that this was the explanation we obtained with our pseudo nonlinear *BOXES* model for the C cycle in Chapter 6 with $\beta = 0.3$ and $\epsilon = 10$.) However, many terrestrial ecologists argue that there is no firm evidence for such a sizable growth in the terrestrial biosphere. On the other hand, some believe that the missing C went into the ocean. However, many marine geochemists argue that the oceans could not have absorbed that much more CO_2. Let's see if we can use our oxygen cycle model to shed some light on this conundrum.

9.7.1. Setting Up the Experiment

We will repeat the same basic numerical experiment we carried out in Chapter 6. Starting with preindustrial conditions at a nominal date of 1860, we will simulate the ensuing 130 years with anthropogenic CO_2 emissions increasing in three steps: (i) From 1860 to 1920, these emissions are set at an annual rate of 1 Gton (or 0.08 kTmole C); (ii) from 1920 to 1960, they are set at 3.5 Gtons C (or 0.29 kTmole C); and (iii) from 1960 to 1990, they are set at 5 Gtons C (or 0.42 kTmole C). This yields a total of 350 Gtons C (or 29 kTmoles C) of excess CO_2 emissions over the 130-year period. As in Chapter 6, we accomplish this in our model by adjusting F_P, the rate of weathering of organic C sediments; that is,

$$r_P = 5.5 \times 10^{-8} \text{ year}^{-1} \qquad \text{for } 0 < t < 60 \text{ years}$$
$$r_P = 1.8 \times 10^{-7} \text{ year}^{-1} \qquad \text{for } 60 < t < 100 \text{ years} \qquad (9.36)$$
$$r_P = 2.55 \times 10^{-7} \text{ year}^{-1} \qquad \text{for } 100 < t < 130 \text{ years}$$

where

$$F_P = r_P C(\text{Org C}) \qquad (\text{kTmoles C year}^{-1}) \qquad (9.37)$$

However, instead of carrying out a single simulation, let's carry out two independent simulations. In Simulation 1, we use the parameters from our Standard Model including $\beta = 0.3$ and $\epsilon = 10$. In Simulation 2, we assume instead that $\beta = 0.0$ and $\epsilon = 5$. (Note that by setting $\beta = 0.0$, we eliminate the CO_2 fertilization effect and thus prevent any of the excess CO_2 from going to the biosphere. On the other hand, by decreasing ϵ to 5, we enhance the uptake of CO_2 by the ocean. As we shall see presently, the two simulation produce the two alternate scenarios for the missing C; in Simulation 1 the missing C ends up in the biosphere and in Simulation 2 the missing C ends up in the ocean.)

9.7.2. The Results

The results of the two simulations are illustrated in Figures 9.6, 9.7, and 9.8. In Figure 9.6, we illustrate the calculated abundance of atmospheric CO_2 for Simulations 1 and 2, and we compare these results with those obtained from our pseudo-nonlinear *BOXES* model and with observations. Interestingly, all three model approaches yield quite similar results and they all agree quite well with observed CO_2 concentrations. Thus we can conclude that, as in the case of the pseudo-nonlinear *BOXES* model, Simulations 1 and 2 of our oxygen cycle model provide a reasonably accurate picture of the evolution of atmospheric CO_2 under the influence of anthropogenic emissions, predicting an airborne fraction of about 44%. Now let's take a look at what our two simulations predict for the fate of the nonairborne fraction.

Figure 9.7 illustrates the fate of the excess C emitted as CO_2 as a function of time for the two simulations. Given the results illustrated in Figure 9.6, it should not be surprising that both simulations show similar amounts of excess C in the at-

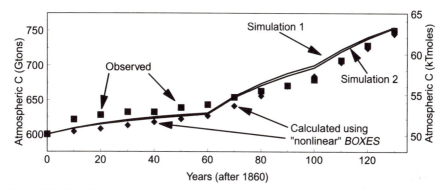

Figure 9.6. Variations in the atmospheric CO_2 reservoir as a function of time since 1860 given a total CO_2 input to the atmosphere by anthropogenic activities of 350 Gtons C. Results are shown for Simulations 1 and 2 using our model for the oxygen cycle and for the pseudo-nonlinear version of *BOXES* from Chapter 6. Reservoir amounts inferred from observed atmospheric CO_2 observations are also shown. Excellent agreement between all four approaches is obtained.

mosphere, with a little less than 10 kTmoles C accumulating there by 1990. However, while Simulation 1 predicts roughly equal amounts of the remaining excess C in the ocean and terrestrial biosphere, Simulation 2 predicts that it all goes to the ocean. Thus, Simulations 1 and 2 provide alternate models for the fate of the missing C; in Simulation 1 the missing C goes to the terrestrial biosphere, while in Simulation 2 it goes to the ocean.

Interestingly, as illustrated in Figure 9.8, the two simulations also predict different effects on the atmospheric O_2 reservoir. Because the uptake of excess CO_2 by the terrestrial biosphere produces atmospheric O_2, while oceanic uptake does not, Simulation 1 predicts a considerably smaller decrease in atmospheric O_2 than that predicted for Simulation 2. While the total effect on atmospheric O_2 is small when compared to the total O_2 reservoir (i.e., less than 0.1%), the difference between the two simulations is nonzero. Highly precise measurements of atmospheric O_2 carried out over a period of years should thus be able to help us to determine the fate of the missing C.[5] And, with this problem solved, we should be in a much better position to predict the effects of future CO_2 emissions and thus the evolution of atmospheric CO_2 in the coming decades.

9.8. CONCLUSION

The oxygen cycle presents us with a complex but wondrous picture of the life support system we call the earth. Through the coupling of one element cycle to another, the earth has evolved into a remarkably stable system. The lesson for the biogeochemist is important and profound: The earth works through complex, interacting

---

[5]In fact, such measurement capabilities now appear to be in hand, and scientists have begun to use this technology to address this important issue. See, for example, Keeling, R. F., and R. Shertz, Seasonal and international variations in atmospheric oxygen and implications for the global carbon cycle. *Nature* **358**, 723–27, 1992.

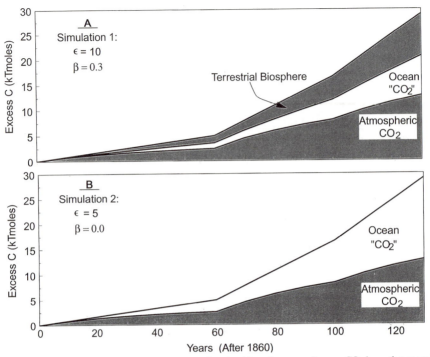

Figure 9.7. The cumulative fate of the excess C emitted into the atmosphere as CO_2 by anthropogenic activities for (A) Simulation 1 and (B) Simulation 2. In Simulation 1, the nonairborne fraction is divided roughly equally between the ocean and the terrestrial biosphere. In Simulation 2, the nonairborne fraction all resides in the ocean. Can you guess what these alternate models for the nonairborne fraction do to atmospheric O_2?

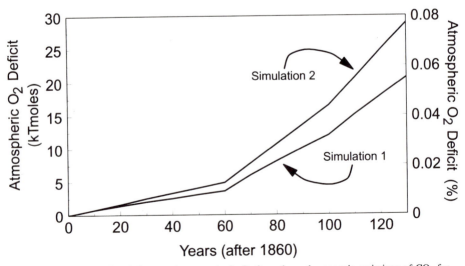

Figure 9.8. The predicted decrease in atmospheric O_2 from the anthropogenic emissions of CO_2 for Simulation 1 and Simulation 2. The results suggest that highly sensitive and precise measurements of atmospheric O_2 could be used to determine if the missing C is going to the ocean (Simulation 2) or to the terrestrial biosphere (Simulation 1).

mechanisms and cycles. Simple models that ignore these interactions and couplings will likely predict a far less stable system than the real one in which we live.

An important inference that we can draw from this conclusion is that our own model, which, after all, only includes the cycles of the five major nutrient elements (P, C, S, N, and O), is also a gross simplification of the earth system and is, no doubt, missing important feedback processes. Other trace elements (e.g., Fe, Si, Hg, Mn, Mg) no doubt have their own roles in the biogeochemical life of our planet. We strongly encourage our readers to consider these other elemental cycles and how they might behave in our changing global environment.

Another lesson is the fact that, because of the complex and interactive nature of the earth system, perturbations in one sector of the system are likely to produce effects in another sector. It is only with a thorough and quantitative understanding of these interactions and their effects that we can understand and ultimately predict their occurrence and size. And then, like the physician studying the laboratory results from a series of blood tests on a patient, we may be able to diagnose the direction our earth system is taking and even, perhaps, devise remedies for a sick planet.

SUGGESTED READING

Garrels, R. M., A. Lerman, and F. T. Mackenzie, Controls of atmospheric O_2 and CO_2: Past, present and future, *American Scientist*, **64**, 306–315, 1976.

Keeling, R. F., and R. Shertz, Seasonal and interannual variations in atmospheric oxygen and implications for the global carbon cycle. *Nature*, **358**, 723–727, 1992.

Van Cappellen, P., and E. D. Ingall, Redox stabilization of the atmosphere and oceans by phosphorus-limited marine productivity, *Science*, **271**, 493–496, 1996.

Walker, J. C. G., Stability of atmospheric O_2, *American Journal of Science*, **274**, 193–214, 1974.

PROBLEMS

1. All the oxygen in the atmosphere is produced at the same time that CO_2 is reduced to organic C. We might therefore expect to find a stoichiometric balance between the amount of O_2 in the atmosphere and the amount of reduced C we find in the earth system. However, inspection of Figures 9.1 and 9.3 reveals no such balance. There are 36,600 kTmoles of O_2 in the atmosphere and more than 1,600,000 kTmoles of reduced C in the sediments. What happened to all the oxygen? Why is it not in the atmosphere?

2. The early 1990s was an unusual time in the life of the earth's greenhouse gases, including CO_2. While anthropogenic activities were believed to have added about 8 Gtons C (or 0.5 kTmoles C) as CO_2 to the atmosphere each year, there was an approximately two-year period in the early 1990s when the atmospheric CO_2 concentration remained essentially flat. In other words, the airborne fraction during the period was 0 and all the CO_2 emitted was absorbed, for some

reason, by the ocean and/or the terrestrial biosphere. The question is: Was it the ocean or was it the biosphere? Suppose you have air samples gathered from the atmosphere before, during, and after this unusual two-year period. Can you devise an experiment involving measurements of O_2 concentrations in these samples to determine which reservoir was responsible for the uptake of the CO_2? How accurate will your measurements need to be in order to make a definitive determination? Are there any other natural processes going on in the atmosphere that might affect the O_2 content of your air samples and thus interfere with your experiment?

Appendix
Equilibrium Constants
at 25°C

A.1. EQUILIBRIUM REACTIONS IN THE H_2O–H_2–O_2 SYSTEM

Equilibrium Reaction	log $K°$
$H^+ + OH^- \rightleftharpoons H_2O$	14.00
$H^+ + e^- \rightleftharpoons {}^1/_2H_2(g)$	0.00
$H^+ + e^- + {}^1/_4O_2(g) \rightleftharpoons {}^1/_2H_2O$	20.78

A.2. EQUILIBRIUM REACTIONS IN THE CO_2–H_2O SYSTEM

Equilibrium Reaction	log $K°$
$CO_2(g) + H_2O \rightleftharpoons H_2CO_3°$	−1.46
$H_2CO_3° \rightleftharpoons H^+ + HCO_3^-$	−6.36
$HCO_3^- \rightleftharpoons H^+ + CO_3^{2-}$	−10.33
$CO_2(g) + H_2O \rightleftharpoons H^+ + HCO_3^-$	−7.82
$CO_2(g) + H_2O \rightleftharpoons 2H^+ + CO_3^{2-}$	−18.15
$CaCO_3$ (calcite) $+ 2H^+ \rightleftharpoons Ca^{2+} + CO_2(g) + H_2O$	9.74

A.3. EQULIBRIUM REACTIONS FOR OTHER CARBON SPECIES

Equilibrium Reaction	log $K°$
$C_6H_{12}O_6°$(glucose) $+ 6H_2O \rightleftharpoons 6CO_2(g) + 24e^- + 24H^+$	5.03

$$C_6H_{12}O_6{}^\circ(\text{glucose}) \rightleftharpoons 6CO(g) + 12e^- + 12H^+ \qquad -16.06$$
$$C_6H_{12}O_6{}^\circ(\text{glucose}) \rightleftharpoons 3CH_3COO^- + 3H^+ \qquad 33.94$$
$$C_6H_{12}O_6{}^\circ(\text{glucose}) \rightleftharpoons 6C(\text{graphite}) + 6H_2O \qquad 89.11$$
$$C_6H_{12}O_6{}^\circ(\text{glucose}) + 24e^- + 24H^+ \rightleftharpoons 6CH_4(g) + 6H_2O \qquad 142.51$$
$$CO(g) + H_2O \rightleftharpoons CO_2(g) + 2e^- + 2H^+ \qquad 3.51$$
$$CH_3COO^- + 2H_2O \rightleftharpoons 2CO_2(g) + 8e^- + 7H^+ \qquad -9.64$$
$$C(\text{graphite}) + 2H_2O \rightleftharpoons CO_2(g) + 4e^- + 4H^+ \qquad -14.01$$
$$CH_4(g) + 2H_2O \rightleftharpoons CO_2(g) + 8e^- + 8H^+ \qquad -22.91$$
$$CO(g) + 6e^- + 6H^+ \rightleftharpoons CH_4(g) + H_2O \qquad 26.43$$
$$CH_3COO^- + 8e^- + 9H^+ \rightleftharpoons 2CH_4(g) + 2H_2O \qquad 36.19$$
$$C(\text{graphite}) + 4e^- + 4H^+ \rightleftharpoons CH_4(g) \qquad 8.90$$

A.4. EQUILIBRIUM REACTIONS FOR NITROGEN SPECIES

Equilibrium Reaction	log K°
$NO_3{}^- \rightleftharpoons NO_3(g) + e^-$	−39.87
$NO_3{}^- + H^+ \rightleftharpoons {}^1/_2N_2O_5(g) + {}^1/_2H_2O$	−9.08
$NO_3{}^- + 2H^+ + e^- \rightleftharpoons NO_2(g) + H_2O$	13.03
$NO_3{}^- + 2H^+ + 2e^- \rightleftharpoons NO_2{}^- + H_2O$	28.64
$NO_3{}^- + 4H^+ + 3e^- \rightleftharpoons NO(g) + 2H_2O$	48.41
$NO_3{}^- + 5H^+ + 4e^- \rightleftharpoons {}^1/_2N_2O^\circ + 5/2H_2O$	75.52
$NO_3{}^- + 6H^+ + 5e^- \rightleftharpoons {}^1/_2N_2(g) + 3H_2O$	105.15
$NO_3{}^- + 10H^+ + 8e^- \rightleftharpoons NH_4{}^+ + 3H_2O$	119.07
$NO_2(g) \rightleftharpoons {}^1/_2N_2O_4(g)$	0.42
$HNO_2{}^\circ \rightleftharpoons H^+ + NO_2{}^-$	−3.15
$NO(g) \rightleftharpoons NO^\circ$	−2.73
$N_2O(g) \rightleftharpoons N_2O^\circ$	0.54
$NH_3(g) \rightleftharpoons NH_3{}^\circ$	1.76
$NH_3{}^\circ + H^+ \rightleftharpoons NH_4{}^+$	9.28

A.5. EQUILIBRIUM REACTIONS OF SULFUR SPECIES

Equilibrium Reaction	log K°
$SO_4{}^{2-} + 2H^+ \rightleftharpoons SO_3(g) + H_2O$	−23.87
$SO_4{}^{2-} + 2e^- + 2H^+ \rightleftharpoons SO_3{}^{2-} + H_2O$	−3.73
$SO_4{}^{2-} + 2e^- + 4H^+ \rightleftharpoons SO_2(g) + 2H_2O$	5.04

$$SO_4{}^{2-} + 2e^- + 4H^+ \rightleftharpoons SO_2{}^\circ + 2H_2O \qquad 5.35$$
$$SO_4{}^{2-} + 6e^- + 8H^+ \rightleftharpoons S_8 + 4H_2O \qquad 35.78$$
$$SO_4{}^{2-} + 7e^- + 8H^+ \rightleftharpoons {}^1/_2S_2{}^{2-} + 4H_2O \qquad 28.54$$
$$SO_4{}^{2-} + 8e^- + 8H^+ \rightleftharpoons S^{2-} + 4H_2O \qquad 20.74$$
$$H_2SO_4{}^\circ \rightleftharpoons H^+ + HSO_4{}^- \qquad 1.98$$
$$HSO_4{}^- \rightleftharpoons H^+ + SO_4{}^{2-} \qquad -1.98$$
$$H_2SO_3{}^\circ \rightleftharpoons H^+ + HSO_3{}^- \qquad -1.91$$
$$HSO_3{}^- \rightleftharpoons H^+ + SO_3{}^{2-} \qquad -7.18$$
$$H_2S(g) \rightleftharpoons H_2S^\circ \qquad -0.99$$
$$H_2S^\circ \rightleftharpoons H^+ + HS^- \qquad -7.02$$
$$HS^- \rightleftharpoons H^+ + S^{2-} \qquad -12.90$$
$$\alpha\text{-FeS(trolite)} \rightleftharpoons Fe^{2+} + S^{2-} \qquad -16.21$$
$$FeS_2(\text{pyrite}) \rightleftharpoons Fe^{2+} + S_2{}^{2-} \qquad -26.93$$
$$CaSO_4 \cdot 2H_2O(\text{gypsum}) \rightleftharpoons Ca^{2+} + SO_4{}^{2-} + 2H_2O \qquad -4.64$$

A.6. EQUILIBRIUM REACTIONS OF PHOSPHORUS SPECIES

Equilibrium Reaction	log K°
$H_3PO_4 \rightleftharpoons H^+ + H_2PO_4{}^-$	2.12
$H_2PO_4{}^- \rightleftharpoons H^+ + HPO_4{}^{2-}$	7.21
$HPO_4{}^{2-} \rightleftharpoons H^+ + PO_4{}^{3-}$	12.67
$H_3PO_3 \rightleftharpoons H^+ + H_2PO_3{}^-$	2.00
$H_2PO_3{}^- \rightleftharpoons H^+ + HPO_3{}^{2-}$	6.59
$H_3PO_4 + 2H^+ + 2e^- \rightleftharpoons H_3PO_3 + 2H_2O$	-9.49
$H_3PO_3 + 3H^+ + 3e^- \rightleftharpoons P + 3H_2O$	-51.36
$P + 3H^+ + 3e^- \rightleftharpoons PH_3$	-2.03

A.7. EQUILIBRIUM REACTIONS OF IRON MINERALS AND COMPLEXES

Equilibrium Reaction	log K°
$Fe(OH)_3(\text{amorp}) + 3H^+ \rightleftharpoons Fe^{3+} + 3H_2O$	3.54
${}^1/_2\alpha\text{-Fe}_2O_3(\text{hematite}) + 3H^+ \rightleftharpoons Fe^{3+} + 3/2H_2O$	0.09
$\alpha\text{-FeOOH(goethite)} + 3H^+ \rightleftharpoons Fe^{3+} + 2H_2O$	-0.02
$Fe^{3+} + H_2O \rightleftharpoons FeOH^{2+} + H^+$	-2.19

$$Fe^{3+} + 2H_2O \rightleftharpoons Fe(OH)_2^+ + 2H^+ \qquad -5.69$$

$$Fe^{3+} + 3H_2O \rightleftharpoons Fe(OH)_3^\circ + 3H^+ \qquad -13.09$$

$$Fe^{3+} + 4H_2O \rightleftharpoons Fe(OH)_4^- + 4H^+ \qquad -21.59$$

$$2Fe^{3+} + 2H_2O \rightleftharpoons Fe_2(OH)_2^{4+} + 2H^+ \qquad -2.90$$

$$Fe^{3+} + H_2PO_4^- \rightleftharpoons FeH_2PO_4^{2+} \qquad 5.43$$

$$Fe^{3+} + H_2PO_4^- \rightleftharpoons FeHPO_4^+ + H^+ \qquad 3.71$$

$$Fe(c) \rightleftharpoons Fe^{2+} + 2e^- \qquad 15.98$$

$$Fe^{3+} + e^- \rightleftharpoons Fe^{2+} \qquad 13.04$$

$$Fe_3O_4(\text{magnetite}) + 8H^+ \rightleftharpoons 3Fe^{3+} + e^- + 4H_2O \qquad -3.42$$

$$FeO(c) + 2H^+ \rightleftharpoons Fe^{2+} + H_2O \qquad 13.48$$

$$Fe(OH)_2(c) + 2H^+ \rightleftharpoons Fe^{2+} + 2H_2O \qquad 12.90$$

$$FeCO_3(\text{siderite}) + 2H^+ \rightleftharpoons Fe^{2+} + CO_2(g) + H_2O \qquad 7.92$$

$$Fe_2SiO_4(\text{fayalite}) + 4H^+ \rightleftharpoons 2Fe^{2+} + H_4SiO_4^\circ \qquad 19.76$$

$$Fe^{2+} + H_2O \rightleftharpoons FeOH^+ + H^+ \qquad -6.74$$

$$Fe^{2+} + 2H_2O \rightleftharpoons Fe(OH)_2^\circ + 2H^+ \qquad -16.04$$

$$Fe^{2+} + 3H_2O \rightleftharpoons Fe(OH)_3^- + 3H^+ \qquad -31.99$$

$$Fe^{2+} + 4H_2O \rightleftharpoons Fe(OH)_4^{2-} + 4H^+ \qquad -46.38$$

$$3Fe^{2+} + 4H_2O \rightleftharpoons Fe_3(OH)_4^{2+} + 4H^+ \qquad -45.39$$

$$Fe^{2+} + H_2PO_4^- \rightleftharpoons FeH_2PO_4^+ \qquad 2.70$$

$$Fe^{2+} + H_2PO_4^- \rightleftharpoons FeHPO_4^\circ + H^+ \qquad -3.60$$

$$Fe^{2+} + SO_4^{2-} \rightleftharpoons FeSO_4^\circ \qquad 2.20$$

Source: Lindsay, W.L., *Chemical Equilibria in Soils*, John Wiley, New York, 1979.

Glossary

Abiotic nitrogen fixation. The conversion of atmospheric nitrogen (N_2) to a form usable by plants (ammonium ion) by an abiological process (i.e., a process not carried out by microbes, plants, etc).

Acid rain. Precipitation having a lower pH than natural rain (5.0–5.6) primarily due to man-made pollutants such as SO_2.

Acid fog. Fog water having a lower pH than natural rain (5.0–5.6) primarily due to man-made pollutants such as SO_2.

Acidity. The quality or state of being acidic.

Activity. A quantity proportional to the concentration of a species in solution, which must be used in the law of mass action to make the law exact.

Adenosine triphosphate (ATP). The primary energy storage molecule in cells. Large amounts of energy are stored in bonds of the three phosphate groups that make up an ATP molecule's tail. The stored energy is released upon hydrolysis and used to drive biochemical processes.

Adenosine diphosphate (ADP). The reaction product of the hydrolysis of ATP (i.e., $ATP + H_2O \rightleftarrows ADP + PO_4^{-3}$)

Airborne fraction. The fraction of carbon dioxide emitted into the atmosphere by human activities that remains in the atmosphere.

Amino acids. A compound containing both an amino group, $-NH_2$, and a carboxyl group, $-COOH$.

Ammonia assimilation. The assimilation of ammonia (or ammonium ion) by plants in the process of photosynthesis.

Ammonia volatilization. The transfer of ammonium ion from soils or natural waters to the atmosphere as gaseous ammonia.

Ammonification. The decomposition of organic matter that leads to the production of ammonia or ammonium compounds, especially by the action of bacteria.

Anabolism. All the metabolic reactions that synthesize complex biomolecules from simpler molecules.

Aqueous-phase species. Species existing as dissolved solutes in an aqueous (i.e., water) solution.

Assimilatory nitrate reduction. The reduction of nitrate to ammonium followed by its assimilation by plants in photosynthesis.

Asthenosphere. The zone of the solid earth that lies beneath the relatively rigid lithosphere. The asthenosphere is considered to be the level of no strain in which there is maximum plasticity and in which the igneous rock magmas are thought to originate.

Autotrophism (autotrophic). The metabolic process of synthesizing organic matter from inorganic matter (e.g., photosynthesis).

Beta factor. The factor that describes the relative enhancement in biospheric net productivity as a result of an increase in atmospheric carbon dioxide concentrations.

Biogeochemical cycles. The various cycles of the elements through the earth's spheres.

Biosphere. The sum total of the living part of the earth system.

Biotic nitrogen fixation. The conversion of atmospheric nitrogen to a form usable by plants (ammonium ion) by organisms (i.e., microbes, plants).

Blue-green algae. Any of a class of algae having the chlorophyll masked by bluish green pigments. These algae have the ability to fix nitrogen.

Brønsted definition of acids and bases. An acid is a proton donor; a base is a proton acceptor.

Catabolism. The destructive metabolism involving the release of energy from the breakdown of complex biomolecules within an organism.

Catalytic cycle. A repeating succession of chemical reactions in which the rate of the reaction is enhanced by a material unchanged chemically at the end of the reaction.

Catalytic destruction cycle. A catalytic cycle in which a chemical species is destroyed.

Chemical energy. The energy stored in a chemical bond.

Chemical thermodynamics. The study of heat and other forms of energy and the various related changes in physical quantities such as temperature, pressure, density, and so on.

Chemoautotrophs. A bacterium capable of oxidizing inorganic compounds to gain energy for its life processes.

Chemoheterotrophs. An organism (e.g., a human being) that ingests organic matter and derives the energy it needs to carry out its metabolic processes through the breakdown of this matter.

Chloroapatite. A constituent of phosphate rock, in which a Cl atom replaces the OH group in hydroxyapatite. (See definition for hydroxyapatite.)

Chloroplast. A plastid that contains chlorophyll and is the site of photosynthesis and starch formation.

Climate sensitivity parameter. The parameter that relates the change in the radiative forcing at the earth's surface to the resulting change in the average surface temperature of the earth.

Colorless bacteria. A group of chemoautotrophic microorganisms lacking in color that are capable of carrying out sulfate reduction to derive energy for the production of organic matter.

Conjugate acid. The species formed when a base accepts a proton.

Conjugate base. The species formed when an acid releases a proton.

Continental margin. One of three major divisions of the ocean basins, being the zones directly adjacent to the continent and including the continental shelf, continental slope, and continental rise.

Continental drift. The slow movement of the continents on a deep-seated viscous zone within the earth.

Continental shelf, slope and rise. The shallow submarine plain of varying width forming a border to a continent ending in a steep slope to the oceanic abyss.

Convergent plate boundary. A boundary between two lithospheric plates that are converging.

Coriolis effect. The apparent force that results when observing objects within a rotating frame of reference. On the earth, the Coriolis force causes projectiles or air currents to move toward the right in the Northern Hemisphere and to the left in the Southern Hemisphere.

Cryosphere. The frozen part of the hydrosphere (i.e., the polar ice caps).

Cycles. The intervals of time during which a sequence of a recurring succession of events or phenomena is completed.

Deep ocean. That part of the ocean lying beneath the ocean's mixed layer.

Degassing. The process by which volatile material trapped within the earth's interior is transferred to the atmosphere.

Delta function. A function which is equal to 1 when the independent variable is 0 but which is equal to 0 everywhere else.

Denitrification. The process by which nitrates or nitrites are reduced and transformed into gaseous nitrogen and nitrous oxide. Most commonly carried out by bacteria.

Divergent plate boundary. A boundary between two lithospheric plates that are moving away from each other.

Earth system. The system of the earth's atmosphere, hydrosphere, lithosphere, and biosphere.

Electronegative. The tendency of an element to attract electrons.

Elementary reactions. A single step in a reaction mechanism.

Enhanced greenhouse effect. The enhanced warming of a planet's surface temperature due to an increased concentration of one or more greenhouse gases (i.e., gases that absorb planetary radiation and re-radiate a portion of this radiation back to the planet's surface).

Euphotic zone. The upper layer of a body of water into which sufficient light penetrates to permit growth of green plants.

Exoergic reactions. A reaction involving a decrease in the total Gibbs free energy.

Exothermic. A reaction that results in the release of heat.

Facultative anaerobic bacteria. A bacterium that can grow with or without the presence of free oxygen.

Fertilization effect. The tendency for plants to increase their net productivity as atmospheric carbon dioxide concentrations increase.

Fixed nitrogen. A nitrogen-containing molecule in which the nitrogen atom is not bound to another nitrogen atom and is thus directly usable by autotrophic organisms.

Fluoroapatite. A major constituent of phosphate rock, in which a F atom replaces the OH group in hydroxyapatite. (See definition for hydroxyapatite.)

Fractionation. The process by which the relative isotopic abundance of an element is enhanced or decreased.

Free energy of formation. A measure of the ability of a chemical system to do useful work

Gaia hypothesis. The hypothesis that the biosphere acts to change and/or stabilize the environment in order to maximize the size or efficiency of the biosphere.

Gas constant. See Universal gas constant.

General circulation. The (climatologically) averaged winds.

Gibbs free energy. A thermodynamic function defined by $G = H - TS$, where H is the enthalpy, T is the thermodynamic temperature, and S is the entropy. It is useful for specifying the conditions of chemical equilibrium for reactions for constant temperature and pressure (G is a minimum).

Global. Relating to the entire world.

Glycine. A sweet crystalline amino acid $C_2H_5NO_2$ obtained by hydrolysis of proteins.

Greenhouse effect. Warming of the earth's surface and the lower layers of atmosphere through the absorption of planetary radiation by so-called greenhouse gases and the re-radiation of a portion of that radiation back to the planet's surface.

Greenhouse gas. A gas that absorbs planetary (i.e., infrared) radiation.

Gross surface ocean production rate (GSOP). The rate of C assimilation via photosynthesis by phytoplankton in the surface ocean.

Gross primary production rate. The rate of carbon assimilation via photosynthesis by green plants on the continents.

Guyout(s). A flat-topped seamount; a former volcano, presumed to have been beveled by wave action and later submerged by crustal subsidence

Henry's law coefficient. The ratio of the partial pressure of a gas and its equilibrium concentration in an aqueous solution.

Heterotrophism. The metabolic process of obtaining energy for cellular processes by taking in food consisting of whole autotrophs or other heterotrophs, their parts, or their waste products.

Hydrosphere. The aqueous envelope of the earth including bodies of water (and the aqueous vapor in the atmosphere).

Hydroxyapatite. A complex phosphate of calcium $Ca_5(PO_4)_3OH$ that occurs as a mineral and is the chief structural element of vertebrate bone.

Igneous. Relating to, resulting from, or suggestive of the intrusion or extrusion of magma or the activity of volcanoes; formed by solidification of molten magma.

Ionic bond. An electrovalent bond; a chemical bond formed between ions of opposite charge.

Lapse rate. The rate of decrease in a meteorological parameter (usually temperature) as a function of height.

Laughing gas. Nitrous oxide.

Linear. Relating to or resembling a straight line, involving a single dimension.

Lithosphere. The solid part of a celestial body (such as the earth) that overlies the asthenosphere; specifically, the outer part of the solid earth and usually considered to be about 50 miles in thickness.

Mafic. Of, relating to, or being a group of usually dark-colored minerals rich in magnesium and iron.

Mesopause. The transition zone between the mesosphere and the thermosphere.

Mesosphere. A layer of the atmosphere extending from the top of the stratosphere to an altitude of about 50 miles.

Metabolic processes. The sum total of the chemical changes in living cells by which energy is provided for vital processes and activities and new material is assimilated.

Metabolic. Relating to, or based on, metabolism (see metabolic processes).

Metamorphic. Relating to metamorphism or a change in the constitution of rock—specifically, a pronounced change effected by pressure, heat, and water that results in a more compact and more highly crystalline condition.

Methane clathrate. A pseudo-frozen form of methane in which methane gas is trapped within the crystal lattice of water. Often found at depth within the ocean.

Mid-ocean ridge(s). One of three major divisions of the ocean basins, being the central belt of submarine mountain topography with a characteristic axial rift. Typically found along a divergent plate boundary.

Mine tailings. The by-products of a mining operation that are left at the mine site to be weathered.

Mineralized. The transformation of an organic compound to inorganic form.

Missing sink. That portion of the nonairborne fraction of carbon dioxide that cannot be accounted for by uptake in the ocean and by the biosphere.

Mitochondria. The organelles of cells that specialize in the harvesting of energy from food molecules and the storage of that energy in ATP.

Net primary production rate. The difference between the gross primary production rate on land (i.e., rate of C assimilation by green plants) and the loss of C from these plants as a result of respiration.

Net surface ocean production rate (NSOP). The rate at which organic C produced within the surface ocean by phytoplankton sinks to the deep ocean.

Neutral rain or precipitation. Rain or precipitation having a pH of 5.6 (i.e., a pH from equilibration with atmospheric CO_2).

Nitrogenase. An enzyme that catalyzes the reduction of gaseous nitrogen to ammonia.

Nitrification. The process by which ammonium ions are oxidized to nitrites and nitrates.

Nitrifying bacteria. Bacteria that carry out nitrification.

Nitrobacteria. Bacteria which catalyze the conversion of nitrite ion to nitrate ion.

Nitrogen fixers. Any of various soil and ocean organisms that fix nitrogen.

Nitrosomonas. Bacteria which catalyze the conversion of ammonium ion to nitrite ion.

Nucleic acid. Any of various acids (as an RNA or a DNA) composed of a sugar or derivative of a sugar, phosphoric acid, and a base and found especially in cell nuclei.

Nucleotides. Any of several compounds that consist of a ribose or deoxyribose sugar joined to a purine or pyrimidine base and to a phosphate group and that are the basic structural units of RNA and DNA.

Null cycle. A chemical cycle of reactions in which the reactants are regenerated as products in the following reactions with no net chemical destruction or production.

Nutrient elements. The elements needed by autotrophic organisms to generate living protoplasm and carry out their metabolic processes.

Ocean upwelling. The process by which deep ocean water is brought to the surface or mixed layer.

Organic polymers. A chemical compound or mixture of compounds formed by the polymerization of organic molecules and consisting essentially of repeating structural units.

Orthophosphates. A compound containing phosphate (i.e., a salt or ester of phosphoric acid)

Oxidation. The loss of one or more electrons from an atom, ion, or molecule.

Oxidation state. For a monatomic ion, the oxidation state is the charge on the ion. For a covalently bonded atom, the oxidation state is the charge on an atom calculated by assigning both electrons of a shared pair to the more electronegative atom. The oxidation state is a formalism, a useful device for counting electrons lost or gained in an oxidation–reduction reaction.

Oxidizing agent. A substance that oxidizes some other species and is itself reduced.

Pangaea. The single continental structure that existed before plate tectonics last caused the continents to split.

Peptide. Any of various amides that are derived from two or more amino acids by combination of the amino group of one acid with the carboxyl group of another and are usually obtained by partial hydrolysis of proteins.

Peptide bond. The chemical bond between carbon and nitrogen in a peptide linkage.

Phosphate ester. A compound formed by the reaction between phosphoric acid and an alcohol, usually with the elimination of water.

Phosphine. A colorless, poisonous, flammable gas, PH_3.

Photoautotrophs (photoautotrophic). Any autotrophic organism that uses radiation (i.e., sunlight) for the energy to generate organic matter.

Photochemical reactions. Chemical reactions involving the interaction of atoms, molecules, or ions with radiant energy.

Photolysis. Chemical decomposition caused by the action of radiant energy.

Photostationary state. A chemical steady-state in the presence of chemically active radiation.

Photosynthesis. The synthesis of chemical compounds with the aid of radiant energy and light; especially formation of carbohydrates in the chlorophyll-containing tissues of plants exposed to light.

Phytoplankton. The passively floating or weakly swimming, usually minute, plant life of a body of water that use radiant energy to synthesize biomass.

Plate tectonics. A concept that envisions the earth's crust divided into various plates that move slowly with respect to each other, being carried along by slow convection currents in the asthenosphere. Along major rifts (such as midocean ridges) the plates are separating and new crust is being created. Elsewhere plates are overriding one another (as at deep ocean trenches) or sliding by one another (as along the San Andreas fault).

Polar covalent bond. A nonionic chemical bond formed by sharing electrons.

Polyprotic acid. A substance with more than one acidic proton.

Primary production. The synthesis and storage of organic molecules during the growth and reproduction of photosynthetic organisms.

Proteins. Any of numerous naturally occurring, extremely complex combinations of amino acids that contain the elements carbon, hydrogen, nitrogen, and oxygen. Proteins are essential constituents of all living cells, and they are synthesized from raw materials by plants but assimilated as separate amino acids by animals.

Purple and green sulfur bacteria. A class of chemoautotrophic bacteria that synthesize organic matter by oxidizing hydrogen sulfide.

Radiative forcing. The net heating of the earth's surface by radiant energy.

Reaction quotient. The same function as the equilibrium constant, but evaluated using any given set of concentrations of the reactants (not necessarily equilibrium concentrations).

Redfield ratio. The approximate proportions of C:N:P in phytoplankton (106:16:1).

Redox. A process involving reduction and oxidation.

Redox equilibria. An equilibrium between an oxidizing agent and a reducing agent and related by electron transfer.

Redox half-reaction. The reaction involving the gain of one of more electrons and the transformation of an oxidized species to a reduced form.

Reducing agent. A chemical species that causes the reduction of another species and is itself oxidized.

Reduction. The process in which a substance gains electrons.

Respiration. The physical and chemical processes by which an organism supplies its cells and tissues with the oxygen needed for metabolism and relieves them of the carbon dioxide formed in energy-producing reaction.

Revelle factor. The factor that relates the relative change in the total concentration of dissolved C in the ocean to a relative change in atmospheric carbon dioxide.

Ridges. A range of hills or mountains. An elongated elevation on an ocean bottom, usually along a divergent plate boundary.

River runoff. The transfer of water and dissolved and particulate matter from the continents to the ocean as a result of the flow of rivers into the ocean.

Rock cycle. The cycle of the formation of sediments at the ocean bottom and their uplift to the earth's surface by plate tectonics and their weathering and erosion.

Scale height. A measure of the rate of decrease in the pressure of a hydrostatic atmosphere.

Sea-floor spreading. The apparent movement or spreading of the sea floor as a result of the divergence of plate boundaries along mid-ocean ridges.

Seamounts. Submarine mountain rising above the deep-sea floor.

Sedimentary. Matter that has sunk to the bottom of a liquid (e.g., the ocean).

Seismic waves. Waves traveling within the lithosphere as a result of vibrations and movements of the lithosphere.

Sialic. Relatively light rock that is rich in silica and alumina and is typical of the outer layers of the earth.

Solubility product. The equilibrium constant that relates the product of the activities of two solutes to the formation of precipitate from these solutes.

Solvation effects. The chemical or physical interactions of a solute ion or molecule with a solvent molecule.

Standard state (of an element). The most stable form of the element at 25°C and 1 atm pressure. A pure solid or a pure liquid is, by definition, in its standard state. A gas is in its standard state if its pressure is 1 atm. A species in solution is in its standard state if its concentration is 1.00 M. Any temperature can be chosen to be the standard state temperature, but in tables of thermodynamic functions the standard state temperature is 25°C.

State variables. A property of a system that has a fixed and definite value for each state of a system. When the state of a system is changed, the values of the change in any state function depends only on the initial and final states of the system and not on the path by which the state of the system is changed.

Stoichiometric reaction. A chemical reaction that may or may not consist of a series of elementary reactions.

Stratopause. The transition region between the stratosphere and the mesosphere.

Stratosphere. The portion of the atmosphere extending from the top of the troposphere (8–15 km) up to about 50 km. The stratosphere is characterized by poor mixing and active photochemistry.

Sulfate-reducing bacteria. Bacteria that utilize sulfate as an oxidant in metabolism of organic matter. The sulfate is concomitantly reduced to sulfide.

Symbiosis. A close association or union of two dissimilar organisms.

Tectonics. Of or relating to the deformation of the earth's crust or the forces involved in or producing such deformation and the resulting forms.

Thermocline. In a thermally stratified body of water, the thermocline separates an upper, warmer, lighter, and oxygen-rich zone from a lower, colder, heavier, oxygen-poor zone; a stratum in which temperature declines at least one degree centigrade with each meter increase in depth.

Thermodynamics. The physics that deals with the mechanical action of relations of heat and energy.

Thermohaline circulation. The water circulation that is dependent upon the cojoint effect of temperature and salinity.

Thermosphere. The part of the earth's atmosphere that begins at about 50 miles above the earth's surface, extends to outer space, and is characterized by steadily increasing temperature with height.

Triphosphate. A salt or acid that contains three phosphate groups and is derived from a complex acid anhydride of orthophosphoric acid.

Tropopause. The transition zone between the troposphere and stratosphere.

Troposphere. The portion of the atmosphere extending from the earth's surface to the tropopause or about 10–12 km. The region is generally characterized by decreasing temperature with altitude, clouds, and active convection.

Turbopause. The region of transition from turbulent mixing to molecular diffusion.

Universal gas constant. The proportionality constant in the ideal gas law, $PV = nRT$. The numerical value of R depends on the units chosen to measure P and V. If P is in atmospheres and V in liters, $R = 0.082057$ liter·atm·mol^{-1} K^{-1}.

Weathering. The action of meteorology and environmental chemistry that alters the color, texture, composition, or form of exposed rocks and ultimately leads to their physical disintegration and chemical decomposition.

Wind-driven circulation. The movement of water at the top of the ocean that is largely driven by frictional forces at the ocean surface from the atmospheric winds.

Index